CISM COURSES AND LECTURES

The series presents lecture notes, monographs, edited works and
proceedings in the field of Mechanics, Engineering, Computer Science
and Applied Mathematics.
Purpose of the series is to make known in the international scientific
and technical community results obtained in some of the activities
organized by CISM, the International Centre for Mechanical Sciences.

INTERNATIONAL CENTRE FOR MECHANICAL SCIENCES

COURSES AND LECTURES - No. 463

FLUID MECHANICS OF SURFACTANT AND POLYMER SOLUTIONS

EDITED BY

VICTOR STAROV
LOUGHBOROUGH UNIVERSITY

IVAN IVANOV
UNIVERSITY OF SOFIA

The publication of this volume was co-sponsored and co-financed by the UNESCO Venice Office - Regional Bureau for Science in Europe (ROSTE) and its content corresponds to a CISM Advanced Course supported by the same UNESCO Regional Bureau.

This volume contains 89 illustrations

In order to make this volume available as economically and as rapidly as possible the authors' typescripts have been reproduced in their original forms. This method unfortunately has its typographical limitations but it is hoped that they in no way distract the reader.

ISBN 978-3-211-21996-6   ISBN 978-3-7091-2766-7 (eBook)
DOI 10.1007/978-3-7091-2766-7

PREFACE

Colloidal systems and dispersions are of great importance in oil recovery, waist water treatment, coating, food and beverage industry, pharmaceutical industry, medicine, environmental protection etc. Colloidal systems and dispersions are always multi-component and multiphase systems. In these systems at least one dimension is in a range of colloidal forces action: colloidal dispersions/emulsions are examples of three dimensional colloidal systems, while thin liquid films are examples of one dimensional colloidal systems. Mostly colloidal systems are stable because their properties are substantially enhanced by the presence of surfactants and or polymers. The distribution and redistribution of the latter molecules is of the crucial importance for colloidal systems.

The contribution presented in this Special Issue deals with flow, distribution and redistribution, coating and deposition of surfactant and polymer molecules in colloidal systems.

In the paper "Effect of Surfactants on Drop Stability and Thin Film Drainage" presented by Professor Krassimir Danov (Sofia University, Bulgaria) the stability of suspensions/emulsions is under consideration. Traditional consideration of colloidal systems is based on inclusion only Van-der-Waals (or dispersion) and electrostatic components, which is refereed to as DLVO (Derjaguin- Landau-Verwey-Overbeek) theory. Professor Danov's contribution shows that not only DLVO components but also other types of the inter-particle forces may play an important role in the stability and colloidal systems. Those contributions are due to hydrodynamic interactions, hydration and hydrophobic forces, steric and depletion forced, oscillatory structural forces. The hydrodynamic and colloidal interactions between drops and bubbles emulsions and foams are even more complex (as compared to that of suspensions of solid particles) due to the fluidity and deformability of those colloidal objects. The latter two features and thin film formation between the colliding particles have a great impact on the hydrodynamic interactions, the magnitude of the disjoining pressure and on the dynamic and thermodynamic stability of such colloidal systems.

In the next paper "Surfactants Effects on Mass Transfer in Liquid-Liquid Systems" Dr Alcina Mendes (Imperial College, UK) reviews the work done by herself and co-workers on the effect of surfactants on mass transfer in binary and ternary liquid-liquid systems. The selected organic-aqueous interfaces has been visualised during the mass transfer process in the presence of ionic and non-ionic surfactants. Results obtained in laboratory and under microgravity conditions are reported. The most significant finding is that surfactants in some cases can induce or increase convection. The latter enhance the mass transfer rate as compared to the Fick's law. The latter means that surfactants can be used to manipulate interfacial stability and particularly in space applications.

The next contribution "Characterisation of Adsorption Layers at Liquid Interfaces – Studies with Drop and Bubble Methods" is presented by Professors Reinhard Miller (Max-Plank-Institute, Germany) and Valentin Fainerman (Medical Physicochemical Center, Ukraine). The non-equilibrium properties of interfacial layers have a large impact on various technologies, comprising food processing, coating, oil recovery, and in particular the formation and stabilisation of foams and emulsions in widespread fields of application. Theoretical models have reached a state that allows a quantitative description of the equilibrium state by thermodynamic models, the adsorption kinetics of surfactants at fluid interfaces, the transfer across interfaces and the response to transient or harmonic perturbations. As result adsorption mechanisms, exchange of matter mechanisms and the dilational rheology are obtained. For some selected surfactant systems, the characteristic parameters obtained on the various levels coincide very well so that a comprehensive understanding was reached.

The next contribution is "Kinetics of Spreading of Surfactant Solutions" by Professor Victor Starov (Loughborough University, UK). Two different mechanism of influence of surfactants on hydrodynamics of spreading are considered: (i) the presence of an inhomogeneous distribution of surfactants on liquid-air interfaces, which results in surface tension gradients, which in their turn cause tangential stresses and Marangoni flow, and (ii) a slow spreading of surfactant solutions over hydrophobic substrates, which is caused by adsorption of surfactant molecules on a bare hydrophobic interface in front of the moving three phase contact line. Results on the theoretical and experimental study of the spreading of an insoluble surfactant over a thin liquid layer are presented, while initial concentrations of surfactant above and below critical micelle concentration (CMC) considered. If the concentration is above CMC two distinct stages of spreading are found (a) the fast first stage, which is connected with the micelles dissolution; (b) the second slower stage, when the surfactant concentration becomes below CMC over the whole liquid surface. During the second stage, the formation of a dry spo in the center of the film is observed. A similarity solution of the corresponding equations for spreading results in good agreement with the experimental observations. Further the spreading of aqueous surfactant solutions over hydrophobic surfaces is considered from both theoretical and experimental points of view. Aqueous droplets do not wet a virgin solid hydrophobic substrate and do not spread, however, surfactant solutions spread. It is shown that the transfer of surfactant molecules from the aqueous droplet onto the hydrophobic surface changes the wetting characteristics in front of the droplet on the moving three phase contact line. The adsorption of surfactant molecules results in an increase of the solid-vapour interfacial tension and hydrophilisation of the initially hydrophobic solid substrate in front of the spreading droplet. This process causes aqueous droplets to spread over time. The time evolution of the spreading of aqueous droplets is predicted and compared with experimental observations. The assumption that surfactant transfer from the droplet surface onto the solid hydrophobic substrate controls the rate of spreading is confirmed by our experimental observations.

The next contribution is "Behaviour of Fluorochemical-Treated Fabric in Contact with Water" by Dr Ana Robert Estelrich (Pymag, S.A., Spain). The introduction of fluorinated coatings in the textile industry has improved significantly the hydro- and oleo phobic properties of the treated fabrics. However, these properties are notably reduced when the fabric is washed and partially recovered with heat treatment in air such as ironing. These changes of repellency are related to the modification of the chemical organisation of the fiber surface. This paper describes the chemical structure of some of the fluoropolymers used in textile and their relationship with these surface properties. A new explanation of the nature of this diminishing performance in contact with water and its recovery by heating is given.

Victor M Starov

CONTENTS

Effect of Surfactants on Drop Stability and Thin Film Drainage

Krassimir D. Danov

Laboratory of Physical Chemistry and Engineering, Faculty of Chemistry, University of Sofia, 1 J. Bourchier Ave., 1164, Sofia, Bulgaria

Abstract. The stability of suspensions/emulsions is under consideration. Traditionally consideration of colloidal systems is based on inclusion only Van-der-Waals (or dispersion) and electrostatic components, which is refereed to as DLVO (Derjaguin-Landau-Verwey-Overbeek) theory. It is shown that not only DLVO components but also other types of the inter-particle forces may play an important role in the stability and colloidal systems. Those contributions are due to hydrodynamic interactions, hydration and hydrophobic forces, steric and depletion forced, oscillatory structural forces. The hydrodynamic and colloidal interactions between drops and bubbles emulsions and foams are even more complex (as compared to that of suspensions of solid particles) due to the fluidity and deformability of those colloidal objects. The latter two features and thin film formation between the colliding particles have a great impact on the hydrodynamic interactions, the magnitude of the disjoining pressure and on the dynamic and thermodynamic stability of such colloidal systems.

1 Introduction

Colloidal systems and dispersions are of great importance in many areas of human activity such as oil recovery, coating, food and beverage industry, cosmetics, medicine, pharmacy, environmental protection etc. They represent multi-component and multiphase (heterogeneous) systems, in which at least one of the phases exists in the form of small (Brownian) or large (non-Brownian) particles (Hetsroni 1982, Russel et al. 1989, Hunter 1993). One possible classification of the colloids is with respect to the type of the continuous phase (dispersions with solid continuous phase like metal alloys, rocks, porous materials, etc. will not be consider).

Gas continuous phase. Examples for liquid-in-gas dispersions are the mist and the clouds. Smoke, dust and some aerosols are typical solid-in-gas dispersions (Britter and Griffiths 1982, Arya 1999).

Liquid continuous phase. Gas-in-liquid dispersions are the foams or the boiling liquids (Prud'homme and Khan 1996, Exerova and Kruglyakov 1998). Liquid-in-liquid dispersions are usually called emulsions. The emulsions exist at room temperature when one of the liquids is immiscible or mutually immiscible in the other, e.g. water, hydrocarbon and fluorocarbon oils and liquid metals (Hg and Ga). Many raw materials and products in food and petroleum industries exist in the form of oil-in-water or water-in-oil emulsions (Shinoda and Friberg 1986, Sjoblom 1996, Binks 1998). The solid-in-liquid dispersions are termed suspensions or sols. The pastes, paints, dyes, some glues and gels are highly concentrated suspensions (Schramm 1996).

The investigations of the *stability of dispersions* against flocculation and coalescence are of crucial importance for the development of new complex fluids. Generally, the stabilizing factors are the repulsive surface forces, the particle thermal motion, the hydrodynamic resistance of medium and the high surface elasticity of fluid particles and films. On the opposite, the factors destabilizing dispersions are the attractive surface forces, the low surface elasticity and gravity, turbulence and other external forces tending to separate the phases. The nonionic and ionic surfactants, polymers and protein blends are widely used in practice to control the surface elasticity and mobility and the magnitude of the surface forces (Adamson and Gast 1997).

The stability of suspensions containing solid particles are treated in the framework of the Derjaguin–Landau–Verwey–Overbeek (DLVO) theory, which accounts for the electrostatic and van der Waals interactions between the particles (Verwey and Overbeek 1948, Derjaguin 1989). In the past decades it has been shown that other types of inter-particle forces may also play an important role in the stability of dispersions – hydrodynamic interactions, hydration and hydrophobic forces, steric and depletion forces, oscillatory structural forces, etc. The hydrodynamic and molecular interactions between surfaces of drops and bubbles in emulsion and foam systems (compared to that of suspensions of solid particles) are more complex due to the particles fluidity and deformability. These two features and the possible thin film formation between the colliding particles have a great impact on the hydrodynamic interactions, the magnitude of the disjoining pressure and on the dynamic and thermodynamic stability of such systems (Ivanov and Dimitrov 1988, Danov et al. 2001, Kralchevsky et al. 2002).

2 Hydrodynamic modeling of surfactant solutions. Interfacial dynamics

A solid (liquid) particle, which moves in a liquid medium, induces a motion of fluid in a hydrodynamic layer around itself and experiences hydrodynamic friction force from the surrounding continuous phase. When the hydrodynamic layers of two colliding particles overlap each other long-range *hydrodynamic interactions* among them due to the viscous friction appear. The quantitative description of this interaction is based on the classical laws of *mass conservation* and *momentum balance* for the bulk phases. If t is time, ∇ is the spatial gradient operator, ρ is the mass density, $\mathbf{v}$ is the local mass average velocity, $\mathbf{P}$ is the hydrodynamic stress tensor, $\mathbf{P_b}$ is the body force tensor and $\mathbf{f} \equiv \nabla \cdot \mathbf{P_b}$ is the body force vector per unit volume acting on the fluid, then the mass and momentum balance equations read (Batchelor 1967, Landau and Lifshitz 1984):

$$\frac{\partial \rho}{\partial t} + \nabla \cdot (\rho \mathbf{v}) = 0 \ , \quad \frac{\partial (\rho \mathbf{v})}{\partial t} + \nabla \cdot (\rho \mathbf{v} \mathbf{v} - \mathbf{P} - \mathbf{P_b}) = 0 \ . \tag{1}$$

For some special kind of magnetic-liquids and liquid crystals the symmetry of the hydrodynamic stress tensor, $\mathbf{P}$, breaks down (Slattery 1978). In the case of classical fluids $\mathbf{P}$ is a symmetric tensor – there is not micro-moments acting on the fluid element. The stress tensor $\mathbf{P}$ becomes a superposition of two contributions – an isotropic thermodynamic pressure, p, and the viscous stress tensor, $\mathbf{T}$, i.e. $\mathbf{P} = -p\mathbf{I}+\mathbf{T}$, where $\mathbf{I}$ is the spatial unit tensor. The viscous stress, $\mathbf{T}$, depends on the fluid velocity gradient characterized by the dilatational, $(\nabla \cdot \mathbf{v})\mathbf{I}$, and the shear rate of strain tensor, $\mathbf{D}-[(\nabla \cdot \mathbf{v})/3]\mathbf{I}$. This dependence is a constitutive law found experimentally for each liquid from bulk rheology experiments. The simplest linear relationship between $\mathbf{T}$ and the rate of strain tensors is referred as the *Newtonian model* of viscosity. The coefficients of proportionality, ξ and η, are called dilatational and shear bulk dynamic viscosity, respectively. Therefore, the Newton law of viscosity reads (Landau and Lifshitz 1984):

$$\mathbf{T} = \xi(\nabla \cdot \mathbf{v})\mathbf{I} + 2\eta[\mathbf{D} - \frac{1}{3}(\nabla \cdot \mathbf{v})\mathbf{I}] \ , \quad \mathbf{D} \equiv \frac{1}{2}[\nabla \mathbf{v} + (\nabla \mathbf{v})^{tr}] \ , \tag{2}$$

where the superscript "tr" denotes conjugation. Pure liquids and surfactant solutions comply well with the Newtonian model (2). Some concentrated macromolecular solutions, foams and emulsions, most of the polymer solutions, colloidal dispersions, gels, etc. exhibit non-Newtonian behavior – for these complex fluids different rheological laws are considered (Barnes et al. 1989).

Applying the constitutive law (2) for motion of homogeneous Newtonian fluids with a constant density, ρ, and shear viscosity, η, to the mass and momentum balance equations (1) one obtains the *Navier-Stokes equation*:

$$\nabla \cdot \mathbf{v} = 0 \ , \quad \rho[\frac{\partial \mathbf{v}}{\partial t} + (\mathbf{v} \cdot \nabla)\mathbf{v}] = -\nabla p + \eta \nabla^2 \mathbf{v} + \mathbf{f} \ . \tag{3}$$

The material derivative d/dt in the left-hand side of Eq. (3) is a sum of a local time derivative, $\partial/\partial t$, and a convective transport derivative, $(\mathbf{v} \cdot \nabla)$. The ratio between the magnitude of convective and viscosity terms is estimated by the Reynolds number. For low shear stresses in the dispersions, the characteristic velocity of the relative particle motion is small enough in order for the Reynolds number to be a small parameter. In this case, the inertia term $\rho d\mathbf{v}/dt$ in Eq. (3) can be neglected. Then, the system of equations becomes linear and the different types of hydrodynamic motions become additive (Happel and Brenner 1965, Russel et al. 1989, Kim and Karrila 1991).

The *kinematics and dynamics boundary conditions* at the interfaces close the hydrodynamic problem (1)-(2). On the solid-liquid boundary the non-slip boundary conditions are applied – the liquid velocity close to the particle boundary is equal to the velocity of particle motion. In the case of pure liquid phases the non-slip boundary condition is replaced by the dynamic boundary condition. The tangential hydrodynamic forces of the contiguous bulk phases, $\mathbf{n} \times (\mathbf{P}+\mathbf{P_b}) \cdot \mathbf{n}$, are equal from both sides of the interface, where $\mathbf{n}$ is the unit normal of the mathematical dividing surface. The capillary pressure compensates the difference between the normal hydrodynamic forces, $\mathbf{n}\cdot(\mathbf{P}+\mathbf{P}_b)\cdot\mathbf{n}$, from both sides of the interface. In both cases the surface between phases is treated as mathematical dividing surface without material properties. In the presence of surfactants the bulk fluid motion near the interface disturbs the homogeneity of the surfactant adsorption distribution. The ensuing surface tension gradients act to restore the interfacial equilibrium. The resulting transfer of adsorbed surfactant molecules, from the regions of lower surface tension toward the regions of higher surface tension, constitutes the Marangoni effect. The analogous effect, for which the surface tension gradient is caused by a temperature gradient, is known as the Marangoni effect of thermocapillarity. In addition, the interfaces possess specific surface rheological properties (surface elasticity and dilatational and shear surface viscosity), which give rise to the so-called Boussinesq effect. Therefore, the surface between phases behaves as a two dimensional material phase with its own physicochemical properties (Slattery 1990, Edwards et al. 1991).

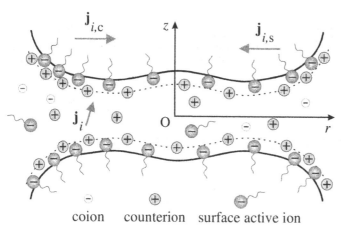

coion   counterion   surface active ion

Figure 1. Schematic picture of a thin liquid film stabilized by ionic surfactant. The ion bulk and surface diffusion fluxes are $\mathbf{j}_i$ and $\mathbf{j}_{i,s}$, respectively. The surface convective flux is $\mathbf{j}_{i,c}$.

To take into account the role of surface-active species the transport equations in the bulk and at surfaces for each of them ($i = 1,2,...,N$) are studied (Dukhin et al. 1995, Danov et al. 1999). In the bulk the change of concentration, c_i, is compensated by the bulk diffusion flux, $\mathbf{j}_i$, bulk convective flux, $c_i\mathbf{v}$, and rate of production due to chemical reactions, r_i (see Fig. 1). The bulk diffusion flux includes the flux driven by external forces (e.g. electro-diffusion), the molecular diffusive and thermodiffusion fluxes. The rate of production, r_i, accounts also for surfactant micellization and micelle decay. Generally, the species can be classified as follows: surface-active ions, which adsorb at the interfaces; counterions, which can adsorb at the so-called Stern layer and coions (Kralchevsky et al. 1999). As a rule the coions do not adsorb (see Fig. 1). The general form of the *species transport equation* in the bulk reads:

$$\frac{\partial c_i}{\partial t} + \nabla\cdot(c_i\mathbf{v}+\mathbf{j}_i) = r_i \ , \quad \text{where } i = 1,2,...,N. \qquad (4)$$

In the two dimensional material phase the change of adsorption, Γ_i, is balanced by the surface convective flux, $\mathbf{j}_{i,c} = \Gamma_i \mathbf{v}_s$, where $\mathbf{v}_s$ is the local material surface velocity, by the surface diffusion flux, $\mathbf{j}_{i,s}$, by the rate of production due to interfacial chemical reactions, $r_{i,s}$, and by the resolved bulk diffusion flux $<\mathbf{n} \cdot \mathbf{j}_i>$, where $<...>$ denotes the difference between the values of a given physical quantity at the two sides of the interface (see Fig. 1). The *surface mass-balance equation* for the adsorption, Γ_i, represents a two dimensional analogue of the bulk transport equation (4):

$$\frac{\partial \Gamma_i}{\partial t} + \nabla_s \cdot (\Gamma_i \mathbf{v}_s + \mathbf{j}_{i,s}) = r_{i,s} + <\mathbf{n} \cdot \mathbf{j}_i > , \qquad (5)$$

where ∇_s is the surface gradient operator (Dukhin et al. 1995, Danov et al. 1999). If a given component k does not adsorb the left-hand side of Eq. (5) is replaced by zero for this species. The interfacial flux, $\mathbf{j}_{i,s}$, contains contributions from the interfacial molecular, electro-, and thermodiffusion. The rate of production, $r_{i,s}$, includes also all possible conformational changes of adsorbed molecules, which lead to the change of the interfacial tension, σ.

Note that the *adsorption isotherms*, relating the surface concentration, Γ_i, with the subsurface value of the bulk concentration, c_i, or the respective kinetic equation for adsorption under barrier control, should also be employed in the computations based on Eqs. (4) and (5) in order to obtain a complete set of equations. These relationships can be found in Dukhin et al. (1995), Danov et al. (1999), Kralchevsky et al. (1999) and in the chapter written by V. B. Fainerman and R. Miller.

In the particular case of molecular electro-diffusion of ionic surfactants (see Fig. 1) the diffusion fluxes for various species ($i = 1,2,...,N$), both amphiphilic and non- amphiphilic with valency Z_i, have the following form (Valkovska and Danov 2001):

$$\mathbf{j}_i = -D_i (\nabla c_i + \frac{Z_i e c_i}{k_B T} \nabla \psi) , \quad \mathbf{j}_{i,s} = -D_i^s (\nabla_s \Gamma_i + \frac{Z_i e \Gamma_i}{k_B T} \nabla_s \psi_s) , \qquad (6)$$

where D_i and D_i^s are the bulk and surface collective diffusion coefficients, respectively, T is the absolute temperature, k_B is the Boltzmann constant, ψ and ψ_s are the bulk and surface electric potential, respectively, and e is the elementary electric charge. The collective diffusion coefficients, D_i and D_i^s, depend on the diffusion coefficients of the individual molecules in the bulk, $D_{i,0}$, and at the surface, $D_{i,0}^s$, on the bulk volume fraction, ϕ_i, on the bulk, μ_i, and surface, μ_i^s, chemical potentials and on the bulk, K_b, and surface, K_s, friction coefficients. The following expressions for D_i and D_i^s are derived in the literature (Stoyanov and Denkov 2001):

$$D_i = \frac{D_{i,0}}{k_B T} \frac{K_b(\phi_i)}{(1-\phi_i)} \frac{\partial \mu_i}{\partial \ln \phi_i} , \quad D_i^s = \frac{D_{i,0}^s}{k_B T} K_s(\Gamma_i) \frac{\partial \mu_i^s}{\partial \ln \Gamma_i} . \qquad (7)$$

The dimensionless coefficient, K_b, accounts for the change in the hydrodynamic friction between the fluid and the particles (created by the hydrodynamic interactions between the particles). The dimensionless surface mobility coefficient, K_s, takes into account the variation of the friction of a molecule in the adsorption layer. The diffusion problem, Eqs. (4) and (5), is connected with the hydrodynamic problem, Eqs. (1) and (2), through the boundary conditions at the material interface.

In contrast with the mathematical dividing surface, which has an isotropic interfacial tension σ, the forces acting on the material interface are not isotropic. They are characterized by the interfacial stress tensor, $\boldsymbol{\sigma}$, which is a two-dimensional counterpart of the bulk stress tensor, $\mathbf{P}$. The two-dimensional analogue of the momentum balance equation (1) written in the bulk is called the *interfacial momentum balance* equation. Note, that the inter-phase exchanges momentum also with the contiguous bulk phases and the corresponding balance equation reads (Slattery 1990, Edwards et al. 1991):

$$\nabla_s \cdot \boldsymbol{\sigma} = \mathbf{n} \cdot < \mathbf{P} + \mathbf{P}_b > . \tag{8}$$

The dependence of the interfacial stress tensor, $\boldsymbol{\sigma}$, on the surface velocity gradients, $\nabla_s \mathbf{v}_s$, and on the physicochemical parameters of the inter-phase is a constitutive law found experimentally for each liquid from interfacial rheology experiments. A number of non-Newtonian interfacial rheological models (Maxwell and Voight type viscoelastic models) have been described in the literature (Edwards et al. 1991, Tambe and Sharma 1991-1994). Following the general principles of non-equilibrium thermodynamics and the Onsager equations one can prove the form of the two-dimensional Newtonian interfacial law (Danov et al. 1997). The interfacial stress tensor contains an isotropic part coming from the thermodynamic adsorption interfacial tension, σ_a, an isotropic part corresponding to the dilatational surface stress, σ_{dil}, and a shear viscous stress, which defines the shear surface viscosity, η_{sh}:

$$\boldsymbol{\sigma} = \sigma_a \mathbf{I}_s + \sigma_{dil} \mathbf{I}_s + \eta_{sh}[\nabla_s \mathbf{v}_s + (\nabla_s \mathbf{v}_s)^{tr} - (\nabla_s \cdot \mathbf{v}_s)\mathbf{I}_s] , \tag{9}$$

where $\mathbf{I}_s$ is the unit surface idemfactor (Slattery 1990, Edwards et al. 1991). In the simplest case when the dilatational surface stress is proportional to the surface deformation with dilatational surface viscosity coefficient η_{dil}, i.e. $\sigma_{dil} = \eta_{dil}(\nabla_s \cdot \mathbf{v}_s)$, equation (9) is called the *Boussinesq-Scriven constitutive law*.

In view of the term $\sigma_a \mathbf{I}_s$ in Eq. (9), the Marangoni effects are hidden in the left-hand side of the interfacial momentum balance equation (8) through the surface gradients of σ_a. The thermodynamic surface tension, σ_a, depends on the adsorption and temperature. The derivatives of σ_a with respect to $\ln \Gamma_i$ and $\ln T$ define the Gibbs elasticity for the i-th surfactant species, E_i, and the thermal analogue of the Gibbs elasticity, E_T:

$$\nabla_s \sigma_a = -\sum_{i=1}^{N} E_i \nabla_s \ln \Gamma_i - E_T \nabla_s \ln T \quad, \quad E_i \equiv -(\frac{\partial \sigma_a}{\partial \Gamma_i})_{T,\Gamma_{j \neq i}} \quad, \quad E_T \equiv -(\frac{\partial \sigma_a}{\partial T})_{\Gamma_i} \quad. \tag{10}$$

The isotropic term $\nabla_s \cdot (\sigma_{dil} \mathbf{I}_s)$ takes into account the role of the disjoining pressure, Π, if the second interface is situated close to the given one (at distance smaller than 200 nm). Other contribution in $\nabla_s \cdot (\sigma_{dil} \mathbf{I}_s)$ is the dilatational surface viscous stress, $\tau_{dil} \mathbf{n}$.

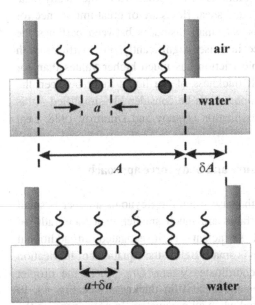

Figure 2. Expansion of a surfactant adsorption layer and definitions of the total, α, and the local, ε, deformations.

To understand the main difficulty in the definition of the surface dilatational viscosity, η_{dil}, let us consider a simple case of uniform expansion of an air/water adsorption layer (see Fig. 2). The surface element with an area A is extended to a new area $A + \delta A$ for a time interval from t to $t + \delta t$ (see Fig. 2). The total deformation, α, and the rate of deformation, $\dot{\alpha}$, are defined as:

$$\dot{\alpha} \equiv \frac{1}{A}\frac{\delta A}{\delta t} \quad, \quad \alpha \equiv \int_0^{\delta t} \dot{\alpha} \, dt \quad. \tag{11}$$

In the case of single surface-active species during the same time, δt, the adsorption changes from Γ to $\Gamma + \delta \Gamma$ and the area per molecule in the adsorption layer – from a to $a + \delta a$ (see Fig. 2). Then, in an analogous way the local deformation, ε, and the rate of local deformation, $\dot{\varepsilon}$, can be introduced:

$$\dot{\varepsilon} \equiv -\frac{1}{\Gamma}\frac{\delta \Gamma}{\delta t} = \frac{1}{a}\frac{\delta a}{\delta t} \quad, \quad \varepsilon \equiv \int_0^{\delta t} \dot{\varepsilon} \, dt \quad. \tag{12}$$

The Boussinesq-Scriven constitutive law postulates that $\tau_{dil} = \eta_{dil} \dot{\alpha}$. The surface viscosity, however, being a property of the adsorption layer itself, should be related to the relative displacement of the adsorbed molecules. As a result the dilatational surface viscous stress should be proportional to the rate of local deformation, i.e. $\tau_{dil} = \eta_{dil} \dot{\varepsilon}$. For insoluble surfactants the both deformations α and ε are equal and the both models are equivalent. For soluble surfactants always the local deformation is smaller than the total deformation, $\varepsilon < \alpha$, because of the bulk diffusion flux (see Fig. 2). The constitutive equation, $\tau_{dil} = \eta_{dil} \dot{\varepsilon}$, suggests that viscous dissipation of energy is possible even at a constant area ($\dot{\alpha}=0$) if the adsorption layer is out of equilibrium ($\dot{\varepsilon}\neq 0$). For surface layer containing a mixture of surfactants the local deformations corresponding to each kind of adsorbed molecules give contribution to the surface dilatational viscous stress and more than one dilatational surface viscosity coefficient appear in the rheological law (Danov et al. 1997).

When the distances between particles are large enough the hydrodynamic interaction between them can be neglected. In this case the solutions of the problems of the steady-state motion of individual droplets (bubbles) in uniform and shear flows are of great importance for the modeling of the bulk viscosity of emulsions. At small distances between particles the magnitude of the hydrodynamic interaction force increases significantly. For particles with tangentially immobile interfaces the hydrodynamic friction has much higher values than for approaching fluid particles. Because of the small thickness of the liquid layer between the interfaces the solution of the problems in the film phase can be considerably simplified – this asymptotic method is called the lubrication approximation (Ivanov and Dimitrov 1988, Leal 1992).

3 Lubrication approximation. Disjoining pressure and body force approach

The *lubrication approximation* can be applied to the case when the Reynolds number is small and when the distances between the particle surfaces are much smaller than their radii of curvature (Reynolds 1886). When the role of surfactants is investigated an additional assumption is made – the Peclet number in the gap is small. Below the equations of lubrication approximation are formulated in a cylindrical coordinate system, Orz, where the droplet interface, S, is defined as $z = H(t,r)/2$ and H is the local film thickness (see Fig. 1). In additional only axial symmetric flows are considered when all parameters do not depend on the meridian angle. The middle plane is $z = 0$ and the unit normal at the surface S pointed to the drop phase is $\mathbf{n}$. The solution in the film (continuous phase) is assumed to be a mixture of nonionic and ionic surfactants and background electrolyte with relative dielectric permittivity ε_f. The general formulation can be found in Kralchevsky et al. (2002).

After integrating Eq. (4) from 0 to $H/2$ along the vertical coordinate z, using the kinematic boundary condition at the interface, and summating the result with Eq. (5) the transport equation of each species is obtained

$$\frac{\partial}{\partial t}\left(\Gamma_i + \int_0^{H/2} c_i \mathrm{d}z\right) = -\frac{1}{r}\frac{\partial}{\partial r}\left\{r\left[\Gamma_i v_{s,r} + j_{is,r} + \int_0^{H/2}(c_i v_r + j_{i,r})\mathrm{d}z\right]\right\} , \quad i = 1,2,\ldots,N. \tag{13}$$

The *integrated mass balance* equation (13) expresses the fact that the local change of the mass of molecules across the film is compensated by the bulk and surface convection and diffusion fluxes. In the case of lubrication approximation for small Peclet numbers the solution of the leading order of the diffusion equations (4) and (6) gives the *Boltzmann type* of the non-equilibrium concentration distribution in the bulk phase:

$$c_i = c_{i,\mathrm{m}} \exp[-\frac{Z_i e}{k_{\mathrm{B}}T}(\psi - \psi_{\mathrm{m}})] \equiv c_{i,\mathrm{n}} \exp(-\frac{Z_i e \psi}{k_{\mathrm{B}}T}) \ , \quad i = 1,2,\ldots,N. \tag{14}$$

The concentration and the electric potential in the middle plane $z = 0$ are $c_{i,\mathrm{m}}(t,r)$ and $\psi_{\mathrm{m}}(t,r)$, respectively, and the concentration $c_{i,\mathrm{n}} \equiv c_{i,\mathrm{m}}\exp[Z_i e\,\psi_{\mathrm{m}}/(k_{\mathrm{B}}T)]$ can be interpreted as the limit of the concentration when the electric potential vanishes. In the case of nonionic surfactant solution ($Z_i e = 0$) $c_{i,\mathrm{n}}(t,r)$ is exactly the concentration of the nonionic components in the film.

The electric potential is related to the bulk charge density through the *Poisson equation* (Landau and Lifshitz 1960). If the Boltzmann type distribution (14) is substituted into the Poisson equation and the result is integrated with respect to z, starting from the middle plane, the following first integral in the lubrication approximation takes place:

$$\frac{\partial^2 \psi}{\partial z^2} = -\frac{e}{\varepsilon_{\mathrm{f}} \varepsilon_0} \sum_{i=1}^{N} Z_i c_i \ , \quad (\frac{\partial \psi}{\partial z})^2 = \frac{2 k_{\mathrm{B}}T}{\varepsilon_{\mathrm{f}} \varepsilon_0} \sum_{i=1}^{N} (c_i - c_{i,\mathrm{m}}) \ . \tag{15}$$

In Eq. (15) ε_0 is the vacuum dielectric constant. The condition for electroneutrality of the solution as a whole is equivalent to the *Gauss law*, which determines the surface charge density, q_{s}. In the lubrication approximation it reads (Kralchevsky et al. 1999):

$$\frac{\partial \psi}{\partial z} = \frac{e}{\varepsilon_{\mathrm{f}} \varepsilon_0} \sum_{i=1}^{N} Z_i \Gamma_i \equiv \frac{q_{\mathrm{s}}}{\varepsilon_{\mathrm{f}} \varepsilon_0} \quad \text{at } z = H(r,t)/2 \ . \tag{16}$$

For ionic surfactant solution the body force tensor, $\mathbf{P}_{\mathrm{b}}$, is not isotropic – it is the Maxwell electric stress tensor, i.e. $\mathbf{P}_{\mathrm{b}} = \varepsilon_{\mathrm{f}} \varepsilon_0 \mathbf{E}\mathbf{E} - \varepsilon_{\mathrm{f}} \varepsilon_0 E^2 \mathbf{I}/2$, where $\mathbf{E} = -\nabla \psi$ is the electric field (Landau and Lifshitz 1960). The density of the electric force plays the role of a spatial body force, $\mathbf{f}$, in the Navier-Stokes equation of motion (3). In the lubrication approximation the pressure in the continuous phase depends on the vertical coordinate, z, only through its *osmotic part* generated from the electric potential and the pressure in the middle plane p_{m} (or the pressure, p_{n}, corresponding to the case of zero potential):

$$p = p_{\mathrm{m}} + k_{\mathrm{B}}T \sum_{i=1}^{N} (c_i - c_{i,\mathrm{m}}) \equiv p_{\mathrm{n}} + k_{\mathrm{B}}T \sum_{i=1}^{N} (c_i - c_{i,\mathrm{n}}) \ . \tag{17}$$

Knowing the pressure distribution, Eq. (17), and the radial component of the surface velocity, $v_{\mathrm{s},r}$, the distribution of the radial component of the bulk velocity is derived from the solution of the Navier-Stokes equation (3) in the lubrication approximation to be (Valkovska and Danov 2001):

$$v_r = v_{s,r} + \frac{4z^2 - H^2}{8\eta}\frac{\partial p_n}{\partial r} + \frac{k_B T}{\eta}\sum_{i=1}^{N}(m_{i2} - m_{i2,s})\frac{\partial c_{i,n}}{\partial r} \quad . \tag{18}$$

The functions, m_{ik}, account for the distribution of i-th ion in the solution and they are defined as

$$m_{i0} \equiv \exp(-\frac{Z_i e \psi}{k_B T}) - 1 \;, \quad m_{ik} \equiv \int_0^z m_{ik-1}dz \quad (i = 1,2,...,N \;\; \text{and} \;\; k = 1,2,3). \tag{19}$$

To close the system of equations for the fluid motion the tangential stress boundary condition and the force balance equation are used. The boundary condition for the balance of the surface excess linear momentum, see equations (8) and (9), takes into account the influence of the surface tension gradient, surface viscosity, and the electric part of the bulk pressure stress tensor. In the lubrication approximation the *tangential stress* boundary condition at the interface, using Eqs. (17) and (18), is simplified to

$$\frac{H}{2}\frac{\partial p_n}{\partial r} + k_B T\sum_{i=1}^{N}m_{1i}\frac{\partial c_{i,n}}{\partial r} + q_s\frac{\partial \psi}{\partial r} + \frac{q_s^2}{2\varepsilon_f\varepsilon_0}\frac{\partial H}{\partial r} = \frac{\partial \sigma_a}{\partial r} + (\eta_{sh} + \eta_{dil})\frac{\partial}{\partial r}[\frac{1}{r}\frac{\partial(rv_{s,r})}{\partial r}] \tag{20}$$

written at $z = H(r,t)/2$. The gradient of the interfacial tension in Eq. (20) is calculated using only the adsorption part of the interfacial tension, σ_a, because the diffuse electric part of σ is already included in the Maxwell electric stress tensor, $\mathbf{P}_b$ (Kralchevsky et al. 1999). The formal limit ($\psi \to 0$) transforms equation (20) into the tangential stress boundary condition for nonionic surfactants widely used in the literature (Ivanov and Dimitrov 1988, Slattery 1990, Singh et al. 1996).

The film between the interfaces thins due to the action of the external force, F_z. For small Reynolds number the external force is balanced by the sum of the hydrodynamic drag force and the intermolecular forces – it is the so-called steady-state approach. In the lubrication approximation from Eq. (17) it follows that:

$$F_z = 2\pi\int_0^\infty(p_m + \Pi_{nel} - p_\infty)rdr = 2\pi\int_0^\infty[p_n + k_B T\sum_{i=1}^{N}(c_{i,m} - c_{i,n}) + \Pi_{nel} - p_\infty]rdr \;, \tag{21}$$

where p_∞ is the pressure at infinity in the meniscus region and Π_{nel} is the disjoining pressure. The disjoining pressure Π_{nel} takes into account all non-electric types of intermolecular interactions, i.e. van der Waals, steric, hydrophobic, oscillatory, etc., except the electrostatic disjoining pressure component (Israelachvili 1992).

There are two ways to take into account the molecular interactions – the first is called the body force approach and the second is known as the disjoining pressure approach. In the case of *body force approach*, the molecular forces are included in the body force, **f**, see Eq. (3). They give contribution into the normal and tangential stress boundary conditions. Felderhof (1968) and Sche and Fijnaut (1976) modeled the van der Waals interactions as an interaction potential bulk force. In the *disjoining pressure approach* the body force, **f**, in the equation of fluid motion (3) is omitted, and the disjoining pressure, Π, is added in the normal stress boundary condition – the hydrodynamic motion does not influences the surface forces. When the body force is potential, i.e. $\mathbf{f} = \nabla U$ and $\mathbf{P_b} = U\mathbf{I}$, the both approaches are equivalent (Maldarelli and Jain 1988). If the body force tensor has anisotropy, e.g. the Maxwell electric stress tensor described above, then the interactions between the film surfaces in dynamic and static conditions are different. The electric part of the disjoining pressure is strictly valid only for static conditions and the electrostatic disjoining pressure described in the next section is applicable for a very slow motion of particles.

4 Surface forces

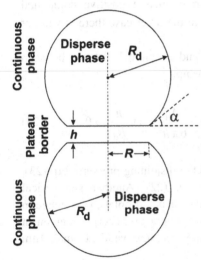

Figure 3. Thin liquid film with radius R between two attached fluid particles.

Thin liquid films can be formed between two colliding emulsion droplets or between the bubbles in foams. The formation of thin films accompanies the particle-particle and particle-wall interactions in colloids. From a mathematical viewpoint a film is thin when its thickness is much smaller than its lateral dimension. From a physical viewpoint a liquid film formed between two macroscopic phases is thin when the energy of interaction between the two phases across the film is not negligible. The specific forces causing the interactions in a thin liquid film are called *surface forces*. The repulsive surface forces stabilize the films and dispersions, whereas the attractive surface forces cause film rupture and coagulation. The molecular theory of surface forces gives expressions for the disjoining pressure corresponding to different types of molecular interactions: van der Waals, electrostatic, steric, oscillatory, hydrophobic, etc. In the recent review by Klitzing and Müller (2002) the factors, which determine the stability of foam films and emulsion films, are considered.

The *van der Waals forces* represent an averaged dipole-dipole interaction, which is a superposition of orientation interactions (between two permanent dipoles, Keesom 1913), induction interaction (between one permanent dipole and one induced dipole, Debye 1920) and dispersion interaction (between two induced dipoles, London 1930). The interaction between two macroscopic bodies depends on the geometry of the system (see Fig. 3). For a plane-parallel film with uniform thickness, h, from component 3 located between two semi-infinite phases composed from components 1 and 2, the disjoining pressure, Π_{VW}, is calculated from the expression (Hamaker 1937):

$$\Pi_{VW} = -\frac{A_H}{6\pi h^3}, \quad A_H = A_{33} + A_{12} - A_{13} - A_{23}. \tag{22}$$

The compound Hamaker constant, A_H, depends on the properties of the phases 1, 2 and 3 through the Hamaker constants, A_{ij}, of hypothetical phase built up from components i and j. If the Hamaker constants A_{ii} and A_{jj} are known the constant A_{ij} can be well approximated as $(A_{ii}A_{jj})^{1/2}$. For films between two symmetric phases A_H is positive and the disjoining pressure corresponds to an attraction between film interfaces. In the case of wetting films the van der Waals disjoining pressure can be positive or negative. For the particular example of aqueous films on polystyrene or polyvinylchloride (PVC) plates we have $A_{PVC/water/air} = -1.15 \times 10^{-20}$ J. The negative value means that the van der Waals forces are repulsive. In contrast, if polytetrafluoroethylene (PTFE) is taken as a solid substrate, then a positive compound Hamaker constant is observed $A_{PFTE/water/air} = +2.65 \times 10^{-21}$ J. In the latter case there is van der Waals attraction.

For two identical spherical particles with radiuses, R_d, and a minimal distance between them, h, the disjoining pressure is calculated from the relationship:

$$\Pi_{VW} = -\frac{32 A_H R_d^5 (16 R_d^2 + 28 R_d h + 7h^2)}{3\pi h^3 (2R_d + h)^4 (4R_d + h)^3} \rightarrow -\frac{A_H}{6\pi h^3} \quad \text{at} \quad \frac{h}{R_d} \rightarrow 0. \tag{23}$$

It is well illustrated that at small distances between particles the disjoining pressure, Eq. (23), obeys the same dependence as for plane-parallel films, Eq. (22). Another geometrical configuration, which corresponds to two colliding deformable emulsion droplets, is sketched in Fig. 3. In this case the exact expression for the van der Waals interaction energy is given in Danov et al. (1993). The asymptotic form for the disjoining pressure valid at small film thickness ($h \ll R_d$) and small film radiuses ($R \ll R_d$) is:

$$\Pi_{VW} = -\frac{A_H}{6\pi h^3}(1 + 3\frac{R^2}{R_d h} - \frac{h}{R_d} - 2\frac{R^2}{R_d^2}) \rightarrow -\frac{A_H}{6\pi h^3}(1 + 3\frac{R^2}{R_d h}) \quad \text{at } h \ll R_d \text{ and } R \ll R_d. \tag{24}$$

Equation (24) demonstrates that the role of Plateau border (see Fig. 3) becomes considerable for calculation of the disjoining pressure for large film radiuses, i.e. $R^2 \gg R_d h$.

Lifshitz (1956) and Dzyaloshinskii et al. (1961) developed an approach to the calculation of the Hamaker constant A_H in condensed phases, called the *macroscopic* theory. The latter is not limited by the assumption for pairwise additivity of the van der Waals interaction. The authors treat each phase as a continuous medium characterized by a given uniform dielectric permittivity, which depends on the frequency, v, of the propagating electromagnetic waves.

For the symmetric configuration of two identical phases i interacting across a medium j the macroscopic theory provides the expression (Russel et al. 1989)

$$A_H \equiv A_{iji} = A_{iji}^{(v=0)} + A_{iji}^{(v>0)} = \frac{3}{4}k_BT\frac{(\varepsilon_i - \varepsilon_j)^2}{(\varepsilon_i + \varepsilon_j)^2} + \frac{3h_Pv_e}{4\pi}\frac{(n_i^2 - n_j^2)^2}{(n_i^2 + n_j^2)^{3/2}}\int_0^\infty \frac{(1 + 2\tilde{h}\,\xi)\exp(-2\tilde{h}\,\xi)}{(1 + 2\xi^2)^2}d\xi \,, \quad (25)$$

where: $h_P = 6.63 \times 10^{-34}$ is the Planck constant; $v_e \approx 3.0 \times 10^{15}$ Hz is the main electronic absorption frequency; n_i and n_j are the refractive indexes, respectively, of the outer phase and inner phase; ε_i and ε_j are the relative dielectric permittivities; the dimensionless thickness $\tilde{h}$ is defined by the expression $\tilde{h} = 2\pi v_e h n_j (n_i^2 + n_j^2)^{1/2}/c_0$; $c_0 = 3.0 \times 10^8$ m/s is the speed of light in vacuum; ξ is an integration variable. The last term in equation (25) takes into account the electromagnetic retardation effect.

The electrostatic interaction between film interfaces becomes operative at distances when the both electric double layers overlap each other. If the particles collide at small velocity of motion the lateral distribution of the ions is approximately uniform and from Eq. (21) an *electrostatic disjoining pressure*, Π_{el}, can be defined:

$$\Pi_{el} = k_BT\sum_{i=1}^N (c_{i,m} - c_{i,\infty}) = k_BT\sum_{i=1}^N c_{i,\infty}[\exp(-\frac{Z_i e \psi_m}{k_BT}) - 1] \,, \quad (26)$$

where $c_{i,\infty}$ is the bulk concentration of the i-th ion. As pointed out by Langmuir (1938) the electrostatic disjoining pressure can be identified with the excess osmotic pressure in the middle plane of the film. The magnitude of Π_{el} depends on the dimensionless distance, κh, where κ is the Debye screening parameter defined with respect to the ionic strength I as:

$$\kappa^2 \equiv \frac{2e^2 I}{\varepsilon_0 \varepsilon_f k_BT} \,, \quad I \equiv \frac{1}{2}\sum_{i=1}^N Z_i^2 c_{i,\infty} \,. \quad (27)$$

To find the exact value of Π_{el} applying Eq. (26) the adsorption isotherms of ions are needed. Combining the isotherms and the Gauss law, Eqs. (15) and (16) written using the surface potential ψ_s in the form

$$\frac{\kappa^2}{4I}(\sum_{i=1}^N Z_i\Gamma_i)^2 = \sum_{i=1}^N c_{i,\infty}[\exp(-\frac{Z_i e \psi_s}{k_BT}) - \exp(-\frac{Z_i e \psi_m}{k_BT})] \,, \quad (28)$$

the value of the electric potential in the middle plane is obtained. A list of simple equations for Π_{el}, valid for different kind of ionic solutions, is reported in Kralchevsky et al. (2002). For particular case of symmetric electrolytes ($Z:Z$ and $c_{1,\infty} = c_{2,\infty} = c_\infty$) and small values of the electric potential in the middle plane one finds (Verwey and Overbeek 1948):

$$\Pi_{el} \approx 64 c_\infty k_B T \tanh^2 (\frac{Z e \psi_s}{4 k_B T}) \exp(-\kappa h) \ . \tag{29}$$

In principle, it is neither possible the surface potential nor the surface charge to be constant. In such case a condition for charge regulation is applied, which in fact represents the condition for dynamic equilibrium of the counterion exchange between the Stern and the diffuse parts of the electric double layer. Contrary to the case of two identically charged surfaces, which always repel each other, the electrostatic interaction between two plane-parallel surfaces of different potentials, ψ_{s1} and ψ_{s2}, or of different charges, σ_{s1} and σ_{s2}, can be either *repulsive or attractive* (Derjaguin et al. 1987). In the case of low surface potentials, when the Poisson-Boltzmann equation can be linearized, the exact expressions for Π_{el} are derived. The formula for the disjoining pressure at constant surface potentials reads:

$$\Pi_{el}^\psi(h) = \frac{\varepsilon_f \varepsilon_0 \kappa^2}{2 \pi} \frac{2 \psi_{s1} \psi_{s2} \cosh(\kappa h) - (\psi_{s1}^2 + \psi_{s2}^2)}{\sinh^2(\kappa h)} \quad \text{at } \psi_{s1} = \text{const. and } \psi_{s2} = \text{const.} \tag{30}$$

and at constant surface charges the following relationship takes place:

$$\Pi_{el}^\sigma(h) = \frac{1}{2 \varepsilon_f \varepsilon_0} \frac{2 \sigma_{s1} \sigma_{s2} \cosh(\kappa h) + (\sigma_{s1}^2 + \sigma_{s2}^2)}{\sinh^2(\kappa h)} \quad \text{at } \sigma_{s1} = \text{const. and } \sigma_{s2} = \text{const.} \tag{31}$$

When the two surface potentials have opposite signs, i.e. $\psi_{s1} \psi_{s2} < 0$, the electrostatic disjoining pressure, Π_{el}^ψ, is negative for all h and corresponds to electrostatic attraction (see Fig. 4a). This result could have been anticipated, since two charges of opposite sign attract each other. More interesting is the case, when $\psi_{s1} \psi_{s2} > 0$, but $\psi_{s1} \neq \psi_{s2}$. In the latter case, the two surfaces repel each other for $h > h_0$, whereas they attract each other for $h < h_0$ (see Fig. 4a) and the electrostatic repulsion has a maximum value. When $\sigma_{s1} \sigma_{s2} > 0$ the electrostatic repulsion takes place, $\Pi_{el}^\sigma > 0$ for all thicknesses h (see Fig. 4b). However, when $\sigma_{s1} \sigma_{s2} < 0$, Π_{el}^σ is repulsive for small thickness, $h < h_0$ and attractive for larger separations. The electrostatic disjoining pressure in this case has a minimum value (see Fig. 4b).

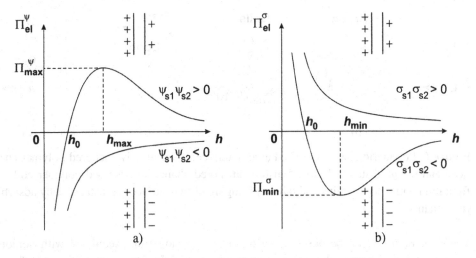

Figure 4. Electrostatic disjoining pressure versus the film thickness, h: a) Π_{el}^{ψ} at fixed surface potential ψ_{s1} and ψ_{s2}; b) Π_{el}^{σ} at fixed surface charges σ_{s1} and σ_{s2}.

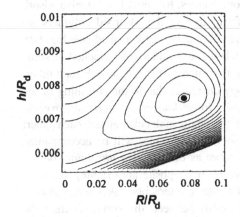

Figure 5. Contour plot of the interaction energy between two deformable charged droplets.

The first quantitative theory of interactions in thin liquid films and dispersions is the DLVO theory. In this theory, the total interaction is supposed to be a superposition of van der Waals and double layer (electrostatic) interactions. The total disjoining pressure is presented in the form, $\Pi = \Pi_{vw} + \Pi_{el}$. A typical disjoining pressure isotherm, Π vs. h, exhibits a maximum representing a barrier against coagulation, and a minimum, called the secondary minimum. With small particles, the depth of the secondary minimum is usually small. If the particles cannot overcome the barrier, coagulation (flocculation) does not take place, and the dispersion is stable due to the electrostatic repulsion, which gives rise to the barrier. With larger colloidal particles the secondary minimum could be deep enough to cause coagulation and even formation of ordered structures of particles. In the case of small Brownian deformable droplets the interaction energy depends on the thickness, h, and on the film radius, R (Fig. 3). The typical contour plot of the interaction energy is plotted in Fig. 5, where the parameters corresponds to micro-emulsions $R_d = 1$ μm, $A_H = 2\times10^{-20}$ J, $\sigma = 1$ mN/m and $\psi_s = 100$ mV. One sees that the minimum of interaction energy (-60 k_BT, the point in Fig. 5) corresponds to a deformed state with equilibrium values of the film radius and thickness (Denkov et al. 1993).

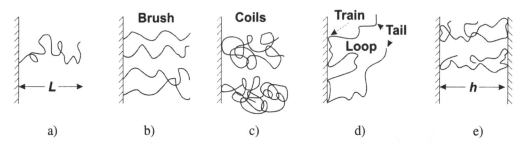

Figure 6. Polymeric chains adsorbed at an interface: a) terminally anchored polymer chain of mean end-to-end distance L; b) a brush of anchored chains; c) adsorbed polymer coils; d) configuration with a loop, trains and tails; e) bridging of two surfaces at distance h by adsorbed polymer chains.

Steric interactions can be observed in foam or emulsion films stabilized with nonionic surfactants or with various polymers, including protein blends. The usual nonionic surfactant molecules are anchored (grafted) to the liquid interface by their hydrophobic moieties. When the surface concentration of adsorbed molecules is high enough, the hydrophilic chains are called to form a brush (see Fig. 6b). The coils of macromolecules, like proteins, can also adsorb at a liquid surface (see Fig. 6c). Sometimes the configurations of the adsorbed polymers are very different from the statistical coil: loops, trains, and tails can be distinguished (see Fig. 6d). The steric interaction between two surfaces appears when chain molecules, attached at some points to a surface, dangle out into the solution. When two such surfaces approach each other, the following three effects take place. First, the entropy decreases due to the confining of the dangling chains which results in a repulsive osmotic force known as *steric* or *overlap* repulsion. Second, in a poor solvent, the segments of the chain molecules attract each other; hence the overlap of the two approaching layers of polymer molecules will be accompanied with some *inter-segment attraction*. Another effect, known as the *bridging attraction*, occurs when two opposite ends of chain molecule can adsorb to the opposite approaching surfaces, thus forming a bridge between them (see Fig. 6e). The steric interaction takes place also in films containing micelles that are pressed (and deformed) between the surfaces when the separation distance is very small.

The steric interaction between two approaching surfaces appears when the film thickness becomes smaller than $2L$, where L is the mean-square end-to-end distance of the hydrophilic portion of the chain. If the chain is entirely extended then L is equal to Nl, with l being the length of a segment and N being the number of segments in the polymer chain. However, due to the Brownian motion $L < Nl$. In the case of a good solvent, the disjoining pressure due to the steric interactions, Π_{st}, can be calculated by means of the Alexander-de Gennes theory as (Alexander 1977, de Gennes 1985 and 1987):

$$\Pi_{st}(h) = k_B T \, \Gamma^{3/2} [(\frac{2L_g}{h})^{9/4} - (\frac{h}{2L_g})^{3/4}] \quad \text{for } h < 2L_g, \quad L_g \equiv N \, \Gamma^{1/3} l^{5/3}, \qquad (32)$$

where Γ is the surface concentration and L_g is the thickness of a brush in a good solvent. The role of the steric interaction on the film stability is reviewed in Klitzing and Muller (2002).

Oscillatory structural forces appear in thin films of pure solvent between two smooth solid surfaces and in thin liquid films containing colloidal particles including macromolecules and surfactant micelles (Israelachvili 1992). In the first case, the oscillatory forces are called the *solvation forces* and they are important for the short-range interactions between solid particles and dispersions. In the second case, the structural forces affect the stability of foam and emulsion films as well as the flocculation processes in various colloids. At lower particle concentrations, the structural forces degenerate into the so-called *depletion attraction*, which is found to destabilize various dispersions.

An illustration of the dependence of the oscillatory disjoining pressure, $\Pi_{osc}(h)$, and its connection with the partial ordering of spheres in the film is shown in Fig. 7. The oscillatory surface forces appear when monodisperse spherical (in some cases, ellipsoidal or cylindrical) particles are confined between two surfaces of a thin film. Even one "hard wall" can induce ordering among the neighboring molecules. The oscillatory structural force is a result of overlap of the structured zones at two approaching surfaces. Rigorous theoretical studies of the phenomenon were carried out by computer simulations and numerical solutions of the integral equations of statistical mechanics. For the sake of estimates, Israelachvili (1992) proposed an analytical expression in which both the oscillatory period and the decay length of the forces were set equal to the particle diameter, d. This oversimplified expression was proved to be unsatisfactory (Kralchevsky and Denkov 1995), as the experimental data with stratifying films give indications of an appreciable dependence of the oscillatory period, d_1, and decay length, d_2, on the particle volume fraction, ϕ. Kralchevsky and Denkov (1995) succeeded in construction a convenient explicit equation for the calculation of the oscillatory structural contribution to the disjoining pressure, Π_{osc}:

$$\Pi_{osc} = P_{osm} \cos(\frac{2\pi h}{d_1}) \exp(\frac{d^3}{d_1^2 d_2} - \frac{h}{d_2}) \quad \text{for } h > d, \quad P_{osm} = k_B T \frac{6\phi}{\pi d^3} \frac{1+\phi+\phi^2-\phi^3}{(1-\phi)^3} . \quad (33)$$

Here P_{osm} is the osmotic pressure in the film interior calculated from the Carnahan-Starling equation. The maximum solid content with three-dimensional close packing of rigid spheres is $\phi_{max} = \pi/(3\times2^{1/2}) \approx 0.7405$ and the oscillatory period, d_1, and the decay length, d_2, are determined by empirical relationships:

$$\frac{d_1}{d} = \sqrt{\frac{2}{3}} + 0.23728\Delta\phi + 0.63300(\Delta\phi)^2 , \quad \frac{d_2}{d} = \frac{0.48663}{\Delta\phi} - 0.42032 , \quad \Delta\phi \equiv \phi_{max} - \phi . \quad (34)$$

Some discussion is needed for the case of ionic surfactants. The charged micelles experience electrostatic interactions and they are not exactly hard spheres. They can be represented as such by taking into account the Debye counterion atmosphere. In this case the effective diameter, $d_{eff} = d_{core} + 2\kappa^{-1}$, is identified with d in equations (33) and (34), where d_{core} denotes the hydrodynamic diameter of the micelles themselves, measured, for instance, by dynamic light scattering (Richetti and Kékicheff 1992, Schmitz 1996). The inverse Debye screening length, κ, is defined by Eq. (27).

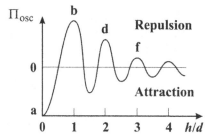

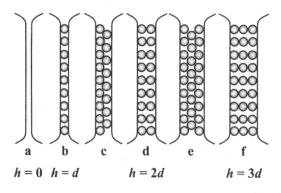

In the case of small Brownian deformable droplets the interaction energy, when the continuous phase is a micellar surfactant solution of sodium nonylphenol polyoxy-ethylene-25 (SNP-25S), is illustrated in Fig. 8 as a function of the thickness, h, and film radius, R (see Fig. 3 for the definition of the geometry). The parameters of the micro-emulsion system are: $R_d = 2$ μm, $d = 9.8$ nm, $\phi = 0.38$, $A_H = 5 \times 10^{-21}$ J, $\sigma = 7.5$ mN/m, $\psi_s = -135$ mV, $\kappa^{-1} = 1.91$ nm, the electrolyte concentration 25 mM. The points on the contour plot (Fig. 8) correspond to tree local minima of -406 k_BT, -140 k_BT and -37 k_BT corresponding to film containing 0, 1 and 2 micellar layers, respectively (Ivanov et al. 1999). These three possible films are thermodynamically stable and they act like barriers against the closer approach and flocculation (or coalescence) of the droplets in emulsions.

Figure 7. Sketch of the consecutive stages of the thinning of the liquid film containing spherical particles. The related oscillatory structural component of the disjoining pressure, Π_{osc}, vs. the film thickness, h, is plotted.

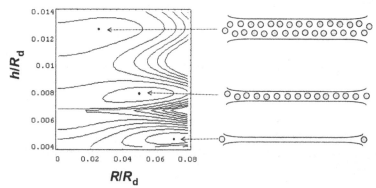

Figure 8. Contour plot of the interaction energy between two oil drops of radius, $R_d = 2$ μm, in the presence of ionic micelles in the water. The parameters correspond to a micellar solution of

sodium nonylphenol polyoxy-ethylene-25 (SNP-25S). The points on the plot correspond to three local minima of 0, 1 and 2 micellar layers, respectively.

The accumulated experimental and theoretical results imply that there are at least two kind of *hydrophobic surface forces* of non-electrostatic physical origin. The first one is due to gaseous capillary bridges (cavitation) between the hydrophobic surfaces (Yushchenko et al. 1983, Pashley et al. 1985, Christenson and Claesson 1988). The formation of gaseous capillary bridges leads to jumps in the experimental force-distance dependence and, moreover, the strength of the interaction increases with the concentration of gas dissolved in water. The second is due to hydrogen-bond-propagated ordering of water molecules in the vicinity of such surfaces (Eriksson et al. 1989, Israelachvili 1992, Paunov et al. 2001). A review was recently published by Christenson and Claesson (2001). In this case the attraction is monotonic and decays exponentially at long distances, with a decay length about 15.8 nm. In principle, the two kinds of hydrophobic surface forces could act simultaneously, and it is not easy to differentiate their effects. If we have a hydrophobic interface, rather than a separate hydrocarbon chain, the ordering of the subsurface water molecules will propagate into the bulk of the aqueous phase. Such ordering is entropically unfavorable. When two hydrophobic surfaces approach each other, the entropically unfavored water is ejected into the bulk, thereby reducing the total free energy of the system. The resulting attraction can in principle explain the hydrophobic surface force of the second kind. For the hydrophobic part of the disjoining pressure, Π_{hb}, Eriksson et al. (1989) derived the following expression:

$$\Pi_{hb}(h) = -\frac{B}{4\pi\lambda}\frac{1}{\sinh^2(h/2\lambda)} . \tag{35}$$

The parameters B and λ characterize, respectively, the strength of the hydrophobic interaction and its decay length. According to Eriksson et al. (1989), B should increase with the degree of hydrophobicity of the surfaces, whereas the decaylength λ should be the same for all hydrophobic surfaces under identical solution conditions. Equation (35) was successfully applied by Paunov et al. (2001) to interpret emulsification data.

Many effects can contribute to the energy of interaction between two fluid particles and the total disjoining pressure, Π, becomes a superposition of all of them, $\Pi = \Pi_{vw} + \Pi_{el} + \Pi_{st} + \Pi_{osc} + \Pi_{hb} + \dots$ (for other contributions in Π see Kralchevsky et al. 2002). For each specified system an estimate may reveal, which of the contributions in the disjoining pressure are predominant, and which of them can be neglected. The analysis shows that the same approach can be applied to describe the multi-droplet interactions in flocs, because in most cases the interaction energy is pair-wise additive.

5　Hydrodynamic interactions in thin liquid films

It is now generally recognized that the presence of surfactants plays an important role for the drainage velocity of thin liquid films and the hydrodynamic forces in these films. The surfactants change not only the disjoining pressure, but also the tangential mobility of the interface of the droplet or bubble. This affects the flocculation and coalescence rate constants, which determine the rates of reversible and irreversible coagulation of the dispersions and emulsions. When two colloidal particles come close to each other, they experience *hydrodynamic forces*, which originate from the interplay of the hydrodynamic flows around two moving colloidal particles or two film surfaces. It becomes important when the separation between the particle surfaces is of the order of the particle radius and increases rapidly with the decrease of the gap width. The simultaneous action of disjoining pressure and hydrodynamic forces determines the stages of the formation and evolution of a liquid film of fluid surfaces (Ivanov et al. 1975, Ivanov and Dimitrov 1988).

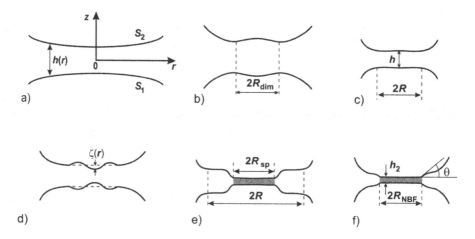

Figure 9. Consecutive stages of evolution of a thin liquid film between two bubbles or drops: a) mutual approach of slightly deformed surfaces; b) the curvature at the film center inverts its sign and a "dimple" arises; c) the dimple disappears and an almost plane-parallel film forms; d) due to thermal fluctuations or other disturbances the film either ruptures or transforms into a thinner Newton black film; e) Newton black film expands; f) the final equilibrium state of the Newton black film is reached.

At large distances the fluid particles approach each other and the hydrodynamic interaction is so small that they are slightly deformed (see Fig. 9a). A dimple is formed at a certain small separation between the fluid particles – the dimple initially grows, but latter becomes unstable and quickly outflows (see Fig. 9b). At a given thickness (called the inversion thickness, h_{it}) a plane-parallel film of radius R and typical thickness from 15 to 200 nm forms (see Fig. 9c). When the long-range repulsive forces are strong enough an equilibrium (but thermodynamically metastable) primary film (called the common black film) can form. The film surface corrugations, caused by thermal fluctuations or other disturbances and amplified by attractive disjoining pressure, may increase their amplitude so much that the film either ruptures or a spot of thinner Newton black film (NBF) forms (see Fig. 9d). If the short-range repulsive disjoining pressure is large enough the black spots (secondary films of very low thickness, $h_2 \approx 5 \div 10$ nm, and a radius R_{sp}) are stable. They either coalescence or grow in diameter, forming an equilibrium secondary (NBF) thin film (see Fig. 9e). After the whole film area is occupied by the Newton black film, the equilibrium between the film and the meniscus is violated and the NBF expands until reaching its final equilibrium radius, R_{NBF}, corresponding to an equilibrium contact angle θ (see Fig. 9f).

The time limiting factor for the approach of two fluid particles determines from the process of mutual approach at large distances (see Fig. 9a) and from the drainage time of the plane-parallel film between deformed interfaces (see Fig. 9c). The solution of the problem about the hydrodynamic interaction between two *rigid spherical particles*, approaching each other across a viscous fluid, was obtained by Taylor. In fact, this solution does not appear in any G.I. Taylor's publications but in the article by Hardy and Bircumshaw (1925) it was published (see Horn et al. 2000). Two spherical emulsion drops of *tangentially immobile* surfaces are hydrodynamically equivalent to the two rigid particles considered by Taylor. The hydrodynamic interaction is due to the viscous dissipation of energy when the liquid is expelled from the gap between the spheres. The geometry of the system is illustrated in Fig. 10. The friction force decreases the approaching velocity of spherical particles, V_{Ta}, proportionally to the decrease of the surface-to-surface distance h in accordance with the equation:

$$V_{Ta} = \frac{2h}{3\pi\eta R_d^2}(F - F_s) , \quad F_s \equiv 2\pi \int_0^\infty \Pi r \, dr , \quad R_d \equiv \frac{2R_{d1}R_{d2}}{R_{d1} + R_{d2}} , \qquad (36)$$

where F is the external force exerted on each drop, F_s is the surface force originating from the intermolecular interactions between the two drops across the liquid medium and R_d is the mean radius of the droplets. Equation (36) is derived using the lubrication approximation and the disjoining pressure approach described in Sec. 3.

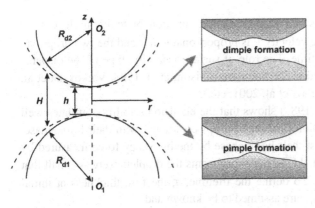

Figure 10. Sketch of two approaching droplets of radii R_{d1} and R_{d2} at a distance h. Possible dimple and pimple formations are illustrated.

When surfactant is present in the continuous phase at not too high concentration, then the surfactant adsorption monolayers, covering the emulsion drops, are tangentially mobile, rather than immobile. The adsorbed surfactant can be dragged along by the fluid flow in the gap between two colliding drops thus affecting the hydrodynamic interaction between them. The appearance of gradients of surfactant adsorption is opposed by the Gibbs elasticity, surface viscosity, surface and bulk diffusion. If the driving force F (say the Brownian or the buoyancy force) is small compared to the capillary pressure of the droplets, the deformation of two spherical droplets upon collision will be only a small perturbation in the zone of contact. Then the film thickness and the pressure within the gap can be presented as a sum of a non-perturbed part and a small perturbation. Solving the resulting hydrodynamic problem for negligible interfacial viscosity, an analytical formula for the velocity of drop approaching, $V = -dh/dt$, is derived for nonionic surfactants (Ivanov and Dimitrov 1988):

$$\frac{V}{V_{Ta}} = \frac{h_s}{2h}[\frac{d+h}{d}\ln(\frac{d+h}{h})-1]^{-1} , \quad h_s \equiv \frac{6\eta D_s}{E_G} , \quad h_a \equiv \frac{\partial \Gamma}{\partial c} , \quad b \equiv \frac{3\eta D}{h_a E_G} , \quad d \equiv \frac{h_s}{1+b} . \quad (37)$$

Here the parameter b accounts for the effect of bulk diffusion, h_s has a dimension of length and takes into account the effect of surface diffusion, E_G is the Gibbs elasticity and h_a is the adsorption length. In the limiting case of very large Gibbs elasticity E_G (tangentially immobile interface) the parameter d tends to zero and then Eq. (37) yields $V \to V_{Ta}$. When the distance between droplets, h, decreases the role of the interfacial mobility increases. The hydrodynamic friction in the gap is much higher that that in the droplets and a small amount of surfactant is enough to prevent the mobility of film interfaces.

In the case without surfactants the energy dissipates also in the drop phases and the viscosity of the disperse phase, η_d, becomes important. A number of solutions, generalizing the Taylor equation (36), have been obtained. In particular, the velocity of central approach of two spherical drops in pure liquid, V, is related to the Taylor velocity, V_{Ta}, through the following expression

$$V = V_{Ta}\frac{1+1.711\xi+0.461\xi^2}{1+0.402\xi} , \quad \xi \equiv \frac{\eta_d}{\eta}\sqrt{\frac{R_d}{2h}} \quad (38)$$

derived by Davis et al. (1989) by means of a Padé-type approximation. Note that in the case of close approach of two drops, $h \to 0$, the velocity V is proportional to $h^{1/2}$ and the two drops can come into contact in a finite period of time. The role of surface viscosity, type of the disperse phase and surfactants are widely studied in the literature (Cristini et al. 1998, Valkovska et al. 1999, Chesters and Bazhlekov 2000, Danov et al. 2001, etc.).

Experimental data (Dickinson et al. 1988) shows that the emulsion stability correlates well with the lifetime of separate thin emulsion films or of drops coalescing with their homophase. The lifetime of oil drops pressed against their homophase by the buoyancy force measured by Dickinson et al. (1988) is plotted in Fig. 11. In the experiments the droplets were so small that the plain-parallel films were not form. To define the lifetime, τ, the film thickness at initial moment, h_{in}, and at the final moment, h_{fin}, are assumed to be known and

$$\tau = \int_{h_{fin}}^{h_{in}} \frac{1}{V} dh \ . \qquad (39)$$

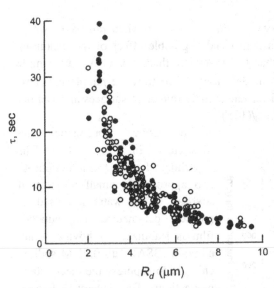

Figure 11. Experimental data of Dickinson et al. (1988) for the lifetime of small drops vs. their radius, R_d.

For the calculation of the velocity, V, in Eq. (39) equations (36) and (37) are used. It is well illustrated that the lifetime is larger for the smaller droplets. The lifetime for slightly deformed droplets decreases with the increase of the driving force, F, and the drop radius, R_d. Note, that for large drops when a plane-parallel film is formed the dependence, τ vs. R_d, is exactly the opposite (see below).

The formation of a dimple (see Fig. 9b) is observed when there is no significant attractive disjoining pressure and the capillary pressure is not large enough to counteract the normal viscous stress and the positive component of the disjoining pressure. Then it may happen that at a given gap width (called *inversion thickness*, h_{it}, Ivanov et al. 1975 and Ivanov and Dimitrov 1988), the droplets deform considerably and their caps becomes flat. Since the flat surface cannot sustain the viscous stress, the interfacial shape in the gap changes suddenly from convex to concave, i.e. a dimple forms. If the disjoining pressure, Π, is negative, the two film surfaces will attract each other, i.e. the disjoining pressure will counteract the hydrodynamic pressure thus decreasing the deformation. Because of the different dependence of these pressures on the thickness, h, (see Sec. 4 where it is shown that Π depends very strong on h), it may happen that at a given thickness, h_{pt}, the disjoining pressure effect totally eliminates the viscous deformation of the caps of the droplets (Fig. 10). At this moment the sum of the dynamic pressure and the disjoining pressure becomes zero. At smaller thickness it will become negative and then protrusions will form at the drop caps. Because of its shape (this protrusions are opposite to the dimple) h_{pt} is called *pimple thickness*. In the case of nonionic surfactants and negligible surface viscosity effect h_{it} and h_{pt} become solutions of the following equations (Valkovska et al. 1999 and Danov et al. 2001):

$$\frac{F - F_s}{2\pi \sigma h_{it}} G(h_{it}) + \frac{R_d}{2\sigma} \Pi(h_{it}) = 1 \ , \quad \frac{F - F_s}{\pi R_d h_{pt}} G(h_{pt}) + \Pi(h_{pt}) = 0 \ , \qquad (40)$$

where the dimensionless mobility function $G(h)$ is defined as

$$G(h) \equiv \frac{d/h - \ln(1 + d/h)}{(1 + d/h)\ln(1 + d/h) - d/h} \cdot \qquad (41)$$

In the case of tangentially immobile interfaces (in Eq. 41 $G \to 1$ when $d/h \to 0$, which corresponds to the limit of a large Gibbs elasticity) and negligible effect of the disjoining pressure one can estimate from Eq. (40) that the inversion thickness obeys the simple expression $h_{it} = F/(2\pi\sigma)$. For mobile surfaces h_{it} decreases. In contrast the pimple thickness does not depend on the interfacial tension and for tangentially immobile surfaces and van der Waals disjoining pressure (see Eq. 23) $h_{pt} = [A_H R_d/(12F)]^{1/2}$.

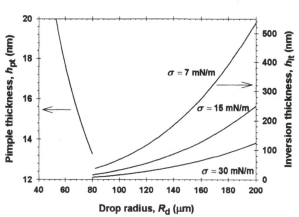

The above conclusions are illustrated in Fig. 12. The experimental data for a system of two approaching small droplets of soybean oil in water are used as physical parameters. The aqueous film is stabilized by bovine serum albumin (BSA) with 0.15 M sodium chloride to suppress the electrostatic interactions. The density difference is 0.072×10^3 kg/m^3, and the experimentally determined interfacial tension was 7, 15 and 30 mN/m for different concentrations of BSA. In the expression for the van der Waals attraction at small and large distances the Hamaker constant

Figure 12. Effect of the drop radius, R_d, on the inversion, h_{it}, and pimple thickness, h_{pt}.

was calculated from Eq. (25). One sees (Fig. 12) that the larger the drop radius the earlier the film forms. The inversion thickness reaches 500 nm for drop radii of 200 µm. The equation (40) for h_{it} has no solution for drop radii smaller than 80 µm. The pimple thickness decreases with the increase of the drop radius, i.e. with the larger driving force. Drops below 80 µm will coalescence or will form a very thin film (NBF) if the steric interaction takes place.

The formation of pimple has been found out by Yanitsios and Davis (1991) in computer calculations for emulsion drops from pure liquids, without any surfactant. Next, by means of numerical calculations Cristini et al. (1998) established the formation of pimple for emulsion drops covered with insoluble surfactant in the case of negligible surface diffusion; their computations showed that a rapid coalescence took part for $h < h_{pt}$. A complete treatment of the problem for the formation of pimple was given by Valkovska et al. (1999), where the effects of surface and bulk diffusion of surfactant, as well as the surface elasticity and viscosity, were taken into account.

In the case when $h < h_{it}$ a liquid film is formed in the zone of contact of two emulsion drops (see Fig. 3 and Fig. 9c). Such configuration appears between drops in flocs and in concentrated emulsions, including creams. In a first approximation, one can assume that the viscous dissipation of energy happens mostly in the thin liquid film intervening between two drops. Some energy dissipates also in the Plateau border. If the interfaces are tangentially immobile (owing to adsorbed surfactant) then the velocity of approach of the two drops can be estimated by means of the *Reynolds formula* for the velocity of approach of two parallel solid discs of radius R, equal to the film radius (Reynolds 1886): $V_{Re} = 2h^3(F - F_s)/(3\pi\eta R^4)$. The Reynolds velocity is used as a scaling factor of the drainage velocity of deformed drops.

When the surfactant is soluble only in the *continuous* phase the respective rate of film thinning V is affected by the surface mobility mainly by through the Gibbs elasticity E_G, just as it is for foam films (Radoev et al. 1974, Traykov et al. 1977 and 1977a, Ivanov 1980, Ivanov and Dimitrov 1988, Danov et al. 1999a). The solution of the lubrication approximation problem for the drainage velocity in the case of nonionic surfactants for the geometry given in Fig. 3 reads (Danov et al. 1999a):

$$\frac{V_{Re}}{V} = \frac{1}{1+b+h_s/h} + \frac{2R_d^2 h^3}{R^4 h_s}\{\frac{R^2}{R_d h} - 1 + [\frac{h}{d}(1 - \frac{R^2}{R_d h}) + 1]\ln(1 + \frac{d}{h})\} . \tag{42}$$

The interfacial parameters b, h_s and d are defined with Eq. (37). Note that the bulk and surface diffusion fluxes, which tend to damp the surface tension gradients and to restore the uniformity of the adsorption monolayers, accelerate the film thinning, see Eqs. (37) and (42). Moreover, since D_s in Eq. (37) is divided by the film thickness h, the effect of surface diffusion dominates that of bulk diffusion for small values of the film thickness. On the other hand, the Gibbs elasticity E_G (the Marangoni effect) decelerates the thinning. Equation (42) predicts that the circulation of liquid in the droplets does not affect the rate of thinning. The role of the Plateau border is well illustrated in Eq. (42). For instance, if the Gibbs elasticity is large enough to prevent the mobility of the film interfaces equation (42) is simplified to:

$$\frac{1}{V} = \frac{1}{V_{Re}} + \frac{1}{V_{Ta}} + \frac{1}{\sqrt{V_{Re}V_{Ta}}} . \tag{43}$$

For $R \to 0$ (non-deformed spherical drops) equation (43) reduces to $V = V_{Ta}$. On the contrary, for $h \to 0$ one has $1/V_{Ta} \ll 1/V_{Re}$, and then Eq. (43) yields $V \to V_{Re}$.

Direct measurements of the lifetime of the drops pressed by buoyancy against a large interface in both Taylor and Reynolds regimes are reported in Basheva et al. (1999) and Gurkov and Basheva (2002) for a wide range of systems. In Fig. 13 the experimental data and the theoretical curve calculated from Eqs. (39) and (43) are presented. The system consists of soybean oil and aqueous solution of 4×10^{-4} wt% BSA + 0.15 M NaCl. The films were detected for droplets above 120 μm in size. Small (micrometer size) droplets are unstable when their radius is larger (Taylor regime). Just the opposite is the case of big drops (above 300 μm) – the lifetime increases with the size (Reynolds regime).

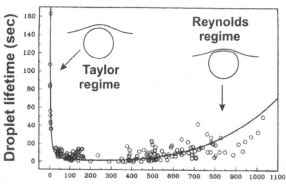

Droplet radius (μm)

Figure 13. Measured lifetime, τ, plotted vs. droplet radius, R_d. Note the existence of shallow and broad minimum. Both Taylor and Reynolds regimes predict very unstable medium-size droplets.

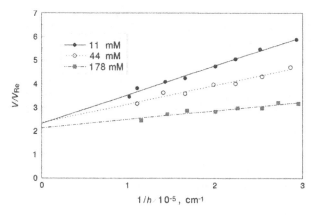

Figure 14. A typical plot of V/V_{Re} versus $1/h$ for nitrobenzene foam films stabilized by various concentrations of dodecanol: 11, 44 and 178 mM.

In the case when the film radius, R, is much larger than $R_d h$ the second term in Eq. (42) is negligible and the Radoev et al. (1974) equation takes place: $V = V_{Re}(1 + b + h_s/h)$. This equation gives a linear dependence of the ratio, V/V_{Re}, on the inverse thickness, $1/h$ (see Fig. 14). From the slope and intersection the values of the bulk and surface diffusion coefficients are determined. The experimental data for nitrobenzene films stabilized by dodecanol (Manev et al. 1977) are processed (Valkovska and Danov 2000). The bulk diffusion coefficient is calculated to be 7.32×10^{-10} m^2/s. The individual surface diffusion coefficient, D_{s0}, is obtained to be a constant 2.90×10^{-9} m^2/s. Note, that the collective surface diffusion coefficient (see Eq. 7) depends on the dodecanol concentration and it changes from 3.55×10^{-9} m^2/s (at 11 mM) to 13.4×10^{-9} m^2/s (at 178 mM). The parameters of the surface tension isotherm at the air/nitrobenzene interface versus the dodecanol concentration are used to determine the adsorption length, h_a, and the Gibbs elasticity, E_G.

It is proved that the role of surface viscosity on the drainage velocity is small (Ivanov and Dimitrov 1988, Singh et al. 1996 and Danov et al. 1999a). In fact, in the dimensionless form of Eq. (20) the surface viscosity term is scaled with the Marangoni term, which is proportional to the Gibbs elasticity. The interfacial properties of all liquids in the presence of surfactants show that if the surface viscosity is not small then the Gibbs elasticity is extremely large and the interfaces become immobile. Therefore, the surface viscosity term can be neglected when the drainage velocity is studied – this term can be important for investigation of the stability of the liquid films.

In the opposite case, when the surfactants are *soluble only in the disperse phase*, the process of convective bulk diffusion is so fast that the liquids behave as pure liquids (Traykov et al. 1977 and 1977a, Ivanov 1980, Ivanov and Dimitrov 1988, Danov et al. 2001). The surfactants remain uniformly distributed throughout the drop surface during the film thinning and interfacial tension gradients do not appear. For that reason, the drainage of the film surfaces is not opposed by surface tension gradients and the rate of film thinning, V, is the same as in the case of pure liquid phases. For an intensive drainage of films with large radii, R, the velocity, V, is calculate from the approximate expression (Ivanov and Dimitrov 1988):

$$\frac{V}{V_{Re}} \approx \frac{1}{\varepsilon_e} \approx \frac{\eta_d \delta}{\eta h} \approx [\frac{108 \pi \eta_d^3 R^4}{\rho \eta h^4 (F - F_s)}]^{1/3} , \tag{44}$$

where ε_e is called emulsion parameter, δ is the thickness of the hydrodynamic boundary layer inside the drops, ρ is the mass density of the continuous phase and η_d is the dynamic shear viscosity of the disperse phase. The validity of Eq. (44) was confirmed experimentally (Traykov et al. 1977a, Ivanov 1980, Ivanov and Dimitrov 1988).

In the beginning stages of mixing in the processes of emulsification large masses of the dispersed phase are embedded in a continuous phase – the large masses of fluids are stretched and folded over. At this stage the capillary number, which is the ration of the viscous forces to the interfacial forces, is very large and the interfacial forces do not play a significant role. As the process evolves, the capillary number decreases and the extended blobs break into many smaller drops. Concurrently, smaller drops begin to collide with each other and may coalescence into larger drops, which may in turn break again. The breakup and coalescence processes complete against each other and it is the result of this competition, which determines the final drop size distribution or morphology. In this case the both phases behave as pure liquids and the drainage velocity of two approaching drops depends considerably from the viscosity of the dispersed and continuous phases and the distance between drops. It leads to a complex hydrodynamic problem, which can be solved only numerically. A list of some approximations is given in Fig. 15 (Chesters 1991 and de Roussel et al. 2001). It is well illustrated that the viscous dissipation of energy in the drops can be important for pure liquids. De Roussel et al. (2001) confirmed experimentally this fact and investigated the mixing of viscous immiscible liquids.

If the thickness of a free liquid film gradually decreases owing to the drainage of liquid, the film typically breaks when it reaches a sufficiently small thickness, called the *critical thickness*, unless some repulsive surface forces are able to provide stabilization (see Fig. 9d – 9f). The mechanism of breakage (in the absence of repulsive forces) was proposed by de Vries (1958) and developed in subsequent studies (Scheludko 1962, Vrij 1966, Ivanov et al. 1974, Ivanov and Dimitrov 1974, Ivanov 1980). According to this mechanism, the film rupture results from the growth of capillary waves at the film surfaces (see Fig. 9d) promoted by attractive surface forces (say, the van der Waals forces), which are operative in the film.

	Drainage rate	Criteria
Rigid drop, $\eta_d \gg \eta$		
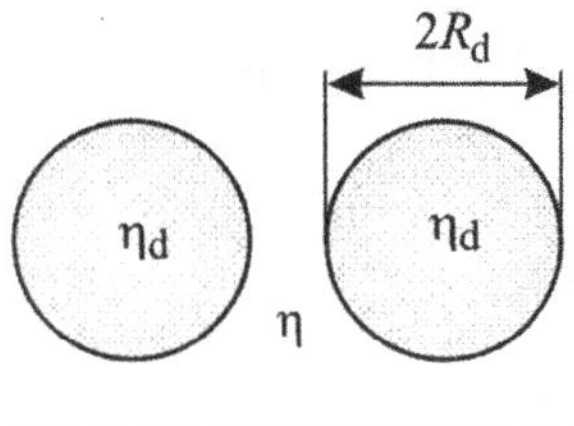	$V \approx \dfrac{2h(F - F_s)}{3\pi \eta R_d^2}$	$h > \dfrac{F - F_s}{2\pi\sigma}$
Immobile interfaces, $\eta_d \gg \eta$		
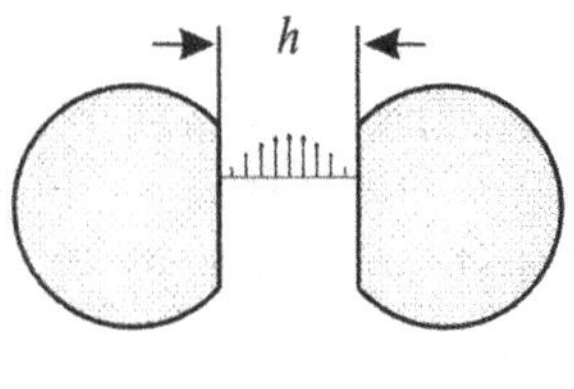	$V \approx \dfrac{8\pi\sigma^2 h^3}{3\eta R_d^2 (F - F_s)}$	$\dfrac{3}{h}[\dfrac{(F - F_s)R_d}{2\pi\sigma}]^{1/2} < \dfrac{\eta_d}{\eta}$
Partially mobile interfaces, $O(\eta) \approx O(\eta_d)$		
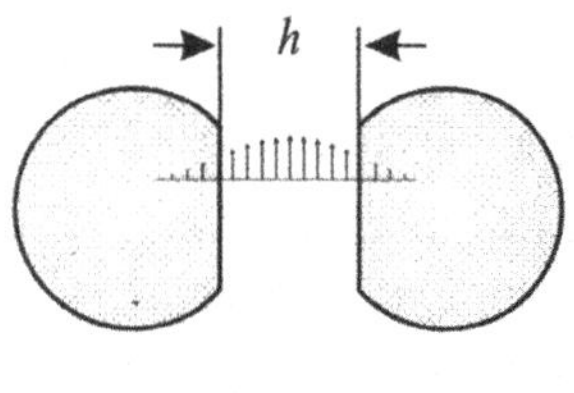	$V \approx \dfrac{2(2\pi\sigma/R_d)^{3/2} h^2}{\pi\eta_d (F - F_s)^{1/2}}$	$6h[\dfrac{2\pi\sigma}{(F - F_s)R_d}]^{1/2} < \dfrac{\eta_d}{\eta} < \dfrac{3}{h}[\dfrac{(F - F_s)R_d}{2\pi\sigma}]^{1/2}$
Fully mobile interfaces, $\eta_d \ll \eta$		
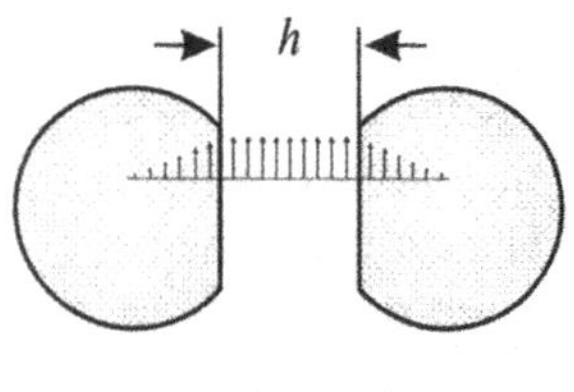	$V \approx \dfrac{2h\sigma}{3\eta R_d}$	$\dfrac{\eta_d}{\eta} < 6h[\dfrac{2\pi\sigma}{(F - F_s)R_d}]^{1/2}$

Figure 15. Rate of drainage of the continuous film between two drops (pure liquids) is determined by the rigidity and mobility of the drops.

In fact, thermally excited fluctuation capillary waves are always present on the film surfaces. With the decrease of the average film thickness, h, the attractive surface force enhances the amplitude of some modes of the fluctuation waves. At the critical thickness, h_{cr}, the two film surfaces locally touch each other due to the surface corrugations, and then the film breaks. A recent version of this capillary-wave model (Danov et al. 2001 and Valkovska et al. 2002), which takes into account all essential physicochemical and hydrodynamic factors, show an excellent agreement with the experiments for the critical thickness of foam and emulsion films (Manev et al. 1984).

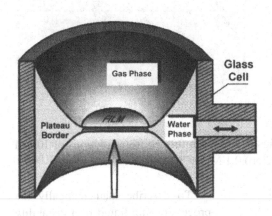

Figure 16. Sketch of the experimental capillary cell.

For experimental observation of the drainage and stability of liquid films the *capillary cell* illustrated in Fig. 16 is widely used (Scheludko and Exerowa 1959). First, the cylindrical glass cell is filled with the working liquid (say, water solution); next, a portion of the liquid is sucked out from the cell through the orifice in the glass wall. Thus, in the central part of the cell a liquid film is formed, which is encircled by a Plateau border. By adjustment of the capillary pressure the film radius, R, is controlled. The arrow (see Fig. 16) denotes the direction of illumination and microscope observation. The optical microscopic observations and the measurements of h are carried out in reflected monochromatic light of wavelength $\lambda_0 = 551$ nm. In particular, h, is determined from the registered intensity, I, of the light reflected from the film, by means of the formula

$$h = \frac{\lambda_0}{2\pi n} \arcsin[\frac{\Delta}{1 + 4Q(1-\Delta)/(1-Q)^2}] , \qquad (45)$$

where $\Delta = (I - I_{min})/(I_{max} - I_{min})$, h is calculated assuming a homogeneous refractive index equal to the bulk solution value; $Q = (n-1)^2/(n+1)^2$; I is the instantaneous value of the reflected intensity, while I_{max} and I_{min} refer to the last interference maximum and minimum values. The intensity, I, is registered by means of a photo-multiplier, whose electric signal is recorded as a function of time.

Typical photos of thinning films are illustrated in Fig. 17. The investigated films behaved in the following way. After the initial dynamic stages of film thinning (see Fig. 9a and 9b), almost plane-parallel films formed (see Fig. 9c), whose thickness was gradually decreasing. At 0.3 M NaCl the film looks dark gray in reflected light just before it ruptures (see Fig. 9d). At 0.1 M NaCl formation of black spots, corresponding to a secondary film, is seen (Fig. 9e). In both photos the film is encircled by Newton interference rings located in the Plateau border.

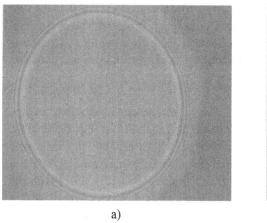 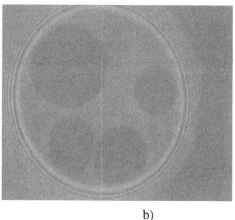

a)   b)

Figure 17. Photos of thinning films of radius R = 155 μm formed from solutions of 10 μM SDS and background electrolyte: a) 0.3 M NaCl; b) 0.1 M NaCl.

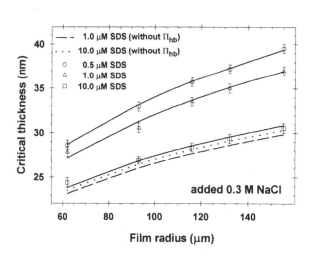

Figure 18. Plot of the critical thickness, h_{cr}, vs. the film radius, R, at 0.3 M fixed concentration of NaCl, for three SDS concentrations, denoted in the figure.

To describe mathematically the process of thin liquid film instability the shape of the corrugated film surfaces is presented as a superposition of Fourier-Bessel modes, proportional to $J_0(kr/R)$, for all possible values of the dimensionless wave number k (J_0 is the zeroth order Bessel function). The mode, which has the greatest amplitude at the moment of film breakage, and which causes the breakage itself, is called the *critical mode*, and its wave number is denoted by k_{cr}. The stability-instability transition for this critical mode happens at an earlier stage of the film evolution, when the film thickness is equal to h_{tr} – the so-called *transitional thickness*, $h_{tr} > h_{cr}$ (Ivanov 1980). The theory provides a system of three equations for determining the three unknown parameters: h_{cr}, k_{cr} and h_{tr}. According to Valkovska et al. 2002, where the most complete theoretical description of the process of simultaneous film drainage and perturbation growth is given, the equations for determining k_{cr} and h_{tr} are:

$$\frac{k_{cr}^2 \sigma}{R^2 h_{cr}^3} \int_{h_{cr}}^{h_{tr}} \frac{h^6}{P_c - \Pi} dh = \int_{h_{cr}}^{h_{tr}} \frac{h^3 \Pi'}{P_c - \Pi} dh \ , \quad \frac{d\Pi}{dh}(h_{tr}) = \frac{24 h_{cr}^3}{h_{tr}^4}[P_c - \Pi(h_{tr})] + \frac{\sigma h_{tr}^3 k_{cr}^2}{2 R^2 h_{cr}^3} \ . \quad (46)$$

The critical thickness, h_{cr}, becomes a solution of the following equation:

$$h_{cr} = (\frac{\sigma h_{tr}^2}{k_B T})^{1/4} h_{tr} \exp[-\frac{k_{cr}^2}{32 h_{cr}^3} \int_{h_{cr}}^{h_{tr}} \frac{h^3}{P_c - \Pi} d\Pi(h)] \ . \quad (47)$$

Note, that the problem (46) and (47) describes the case of tangentially immobile interfaces. One proves that the mobility of interfaces changes only slightly the values of h_{cr} – it affects the value of the critical wave number, k_{cr}.

The data points in Fig. 18 are processed using equations (46) and (47). The disjoining pressure is presented as a sum of the van der Waals attraction due to Eqs. (22) and (25) and the hydrophobic attraction described by Eq. (35). All parameters for the calculation of Π_{vw} are known. The only adjustable parameters are B and λ, which characterize the hydrophobic interaction, Eq. (35). As B characterizes the interfacial hydrophobicity, while λ is a bulk property related to the propagating hydrogen-bonding of water molecules, from the fit of the data three different values of B (denoted by B_1, B_2 and B_3) for the three experimental SDS concentrations, and a single value of λ, the same for the whole set of data are obtained. Note, that the used SDS concentrations are extremely small and they cannot affect the hydrogen bonding in the bulk (and the value of λ), while the adsorption of SDS is material and can affect the interfacial hydrophobicity (and the value of B). The results are: $B_1 = 6.56 \times 10^{-4}$ (for 0.5 µM SDS + 0.3 M NaCl); $B_2 = 4.71 \times 10^{-4}$ (for 1.0 µM SDS + 0.3 M NaCl); $B_3 = 3.34 \times 10^{-5}$ (for 10 µM SDS + 0.3 M NaCl) and $\lambda = 15.85$ nm. The determined decay length of the hydrophobic force in foam films, $\lambda = 15.85$ nm, practically coincides with the value $\lambda = 15.8$ obtained in Eriksson et al. (1989) for mica covered with hydrocarbon monolayer (DDOA = dimethyl-dioctadecyl-ammonium bromide) and with monolayer from fluorinated cationic surfactant. The lower two curves in Fig. 18, are calculated substituting $B = 0$, which means that the hydrophobic surface force is set zero ($\Pi_{hb} = 0$). In this case, all parameters of the theory are known, and the theoretical curves are drawn without using any adjustable parameters. The respective computed curve for 1 µM SDS practically coincides with that for 0.5 µM SDS in Fig. 18. The curves calculated for $\Pi_{hb} = 0$ lay far away from the experimental points for 0.5 and 1 µM SDS – this fact can be interpreted as a consequence of the action of the hydrophobic force in the film. Note also that $h_{cr}(10 \text{ µM SDS}) > h_{cr}(1 \text{ µM SDS})$ for the curves calculated assuming $\Pi_{hb} = 0$, whereas exactly the opposite tendency holds for the respective experimental points.

6 Concluding remarks

There have been numerous attempts to formulate simple rules connecting the emulsion stability with the surfactant properties. Historically, the first one was the Bancroft rule (1913), which states that "in order to have a stable emulsion the surfactant must be soluble in the continuous phase". A more sophisticated criterion was proposed by Griffin (1954) who introduced the concept of hydrophilic-lipophilic balance (HLB). As far as emulsification is concerned, surfactants with an HLB number in the range 3 to 6 must form water-in-oil (W/O) emulsions, whereas those with HLB numbers from 8 to 18 are expected to form oil-in-water (O/W) emulsions. Different formulae for calculating the HLB numbers are available; for example, the Davies' expression (1957) reads: HLB = 7 + (hydrophilic group number) − $0.475n_g$, where n_g is the number of $-CH_2-$ groups in the lipophilic part of the molecule. Schinoda and Friberg (1986) proved that the HLB number is not a property of the surfactant molecules only, but it also depends strongly on the temperature (for nonionic surfactants), on the type and concentration of added electrolytes, on the type of oil phase, etc. They proposed using the phase inversion temperature (PIT) instead of HLB for characterization of the emulsion stability. Davis (1993) summarized the concepts about HLB, PIT and Windsor's ternary phase diagrams for the case of microemulsions and reported topological ordered models connected with the Helfrich membrane bending energy.

Ivanov (1980) have proposed a semi-quantitative theoretical approach that provides a straightforward explanation of the Bancroft rule for emulsions. This approach is based on the idea of Davies and Rideal (1963) that both types of emulsions are formed during the homogenization process, but only the one with lower coalescence rate survives. If the initial drop concentration for the two emulsions (System I and II, see Figs. 19 and 20) is the same, the corresponding coalescence rates for the two emulsions will be proportional to the respective velocities of film thinning, V_I and V_{II} [163]: Rate $_I$/Rate$_{II}$ ≈ V_{II}/V_I. In the case of *deforming drops*, using Eqs. (42) and (44) one derives:

$$\frac{\text{Rate I}}{\text{Rate II}} \approx (486\rho_2 D_{1s}^3)^{1/3} (\frac{h_{cr,I}^3}{h_{cr,II}^2})^{1/3} (\frac{\eta_2}{R^2})^{1/3} \frac{2\sigma/R_d - \Pi_I}{E_G (2\sigma/R_d - \Pi_{II})^{2/3}} , \qquad (48)$$

where $h_{cr,I}$ and $h_{cr,II}$ denote the critical thickness of film rupture for the two emulsion systems in Fig. 19, Π_I and Π_{II} denote the disjoining pressure of the respective films and the indexes 1 and 2 corresponds to the phase 1 and 2, respectively. The product of the first three multipliers in the right-hand side of Eq. (48), which are related to the *hydrodynamic* stability, is c.a. 8×10^{-5} $dyn^{2/3}cm^{-1/3}$ for typical parameter values (Ivanov and Kralchevsky 1997). The last multiplier in Eq. (48) accounts for the *thermodynamic* stability of the two types of emulsion films. Many conclusions can be drawn from Eq. (48).

In thick films the disjoining pressures, Π_I and Π_{II}, are zero, and then the ratio in Eq. (48) will be very small. Consequently, emulsion I (surfactant soluble in the continuous phase) will coalesce much more slowly than emulsion II; hence emulsion I will survive. Thus we get an explanation of the empirical Bancroft rule. The emulsion behavior in this case is controlled mostly by the hydrodynamic factors, i.e. the factors related to the *kinetic* stability.

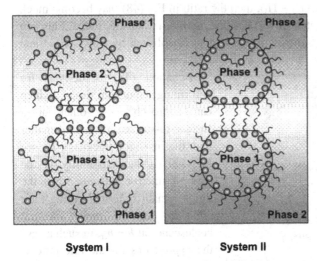

Figure 19. Two possible types of emulsions obtained just after homogenization for deformed drops: the surfactants are soluble in phase 1.

The Gibbs elasticity, E_G, favors the formation of emulsion I, because it slows down the film thinning. On the other hand, increased surface diffusivity, D_{1s}, decreases this effect, because it helps the interfacial tension gradients to relax thus facilitating the formation of emulsion II.

The film radius, R, increases, whereas the capillary pressure, $2\sigma/R_d$, decreases with the rise of the drop radius. Therefore, larger drops will tend to form emulsion I, although the effect is not very pronounced, see Eq. (48). The difference between the critical thicknesses of the two emulsions affects only slightly the rate ratio.

The viscosity of the surfactant-containing phase, η_1, does not appear in Eq. (48) and there is only a weak dependence on η_2. This fact is consonant with the experimental findings about a negligible effect of viscosity (Davies and Rideal 1963, p. 381). The interfacial tension, σ, affects directly the rate ratio through the capillary pressure. The addition of electrolyte would affect mostly the electrostatic component of the disjoining pressure, see Eq. (26), which is suppressed by the electrolyte – the latter has a destabilizing effect on O/W emulsions.

Surface-active additives (such as cosurfactants, demulsifiers, etc.) may affect the emulsifier partitioning between the phases and its adsorption, thereby changing the Gibbs elasticity and the interfacial tension. The surface-active additive may change also the surface charge (mainly through increasing the spacing among the emulsifier ionic headgroups) thus decreasing the electrostatic disjoining pressure and favoring the W/O emulsion. Polymeric surfactants and adsorbed proteins increase the steric repulsion between the film surfaces; they may favor either of the emulsions O/W or W/O depending on their conformation at the interface and their surface activity (Danov et al. 2001, Kralchevsky et al. 2002). The temperature affects strongly both the solubility and the surface activity of non-ionic surfactants (Adamson and Gast 1997). It is well known that at higher temperature non-ionic surfactants become more oil-soluble, which favors the W/O emulsion. These effects may change the type of emulsion formed at the phase inversion temperature (PIT). The temperature effect has numerous implications, some of them being the change of the Gibbs elasticity, E_G, and the interfacial tension, σ.

The disjoining pressure, Π, can substantially change, and even reverse, the behavior of the system if it is comparable by magnitude with the capillary pressure, $2\sigma/R_d$. For instance, if $(2\sigma/R_d - \Pi_{II}) \to 0$ at finite value of $(2\sigma/R_d - \Pi_I)$, then the ratio in Eq. (48) may become much larger than unity, which means that System II will become *thermodynamically* stable. This fact can explain some exclusion from the Bancroft rule, like that established by Binks (1993 and 1998).

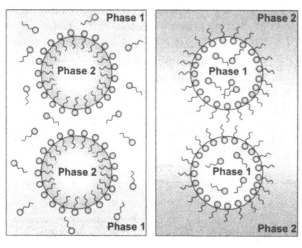

Figure 20. Two possible types of mini-emulsions obtained just after homogenization for spherical drops: the surfactant is soluble in phase 1.

Equation (48) is derived for deforming emulsion drops, which can approach each other at a surface-to-surface distance smaller than the inversion thickness, h_{it}, see Eq. (40). Other possibility is the case of micro-emulsions when the drops remain *spherical* during their collision, up to their eventual coalescence at $h = h_{cr}$. In such a case the expressions for the rate ratio is different.

Let us first consider the case of System II (surfactant inside the drops, Fig. 20) in which case the two drops approach each other like drops from pure liquid phases (if only the surface viscosity effect is negligible). Therefore, to estimate the velocity of approach of such two aqueous droplets one can use equation (38). On the other hand, the velocity V_I of droplet approach in System I can be expressed by means of Eq. (37). Note that the Taylor velocities for System I and System II are different because of differences in viscosity and droplet-droplet interaction. Then the following criterion for formation of emulsion of type I and II is obtained (Ivanov and Kralchevsky 1997 and Danov et al. 2001):

$$\frac{\text{Rate I}}{\text{Rate II}} \approx 1.233 \frac{h_s}{2h} \left(\frac{h}{R_d}\right)^{1/2} \left[\frac{d+h}{d}\ln\left(\frac{d+h}{h}\right) - 1\right]^{-1} \frac{(F - F_s)_I}{(F - F_s)_{II}} . \qquad (49)$$

For typical emulsion systems one has $R_d \gg h_{cr}$, and equation (49) yields Rate I/Rate II $\ll 1$. Therefore, System I (with surfactant in the continuous phase, Fig. 20) will survive. This prediction for spherical drops is analogous to the conclusion for deformable drops. Both these predictions essentially coincide with the Bancroft rule and are valid for cases, in which the *hydrodynamic* stability factors prevail over the *thermodynamic* ones. The latter become significant close to the equilibrium state, $F_s \approx F$, and could bring about exclusions from the Bancroft rule, especially when $(F - F_s)_{II} \to 0$.

The increase of the bulk and surface diffusivities, D_1 and D_{1s}, which tend to damp the surface tension gradients, leads to an increase of the parameters b and d (see Eq. 37), which decreases the difference between Rate I and Rate II. In contrast, the increase of the Gibbs elasticity, E_G, leads to decrease of d (see Eq. 37) and thus favors the survival of System I. These are the same tendencies as for deforming drops.

Equations (48) and (49) lead to some more general conclusions than the original Bancroft rule (e.g. the possibility for inversion of the emulsion stability due to disjoining pressure effects). We neither claim that Bancroft rule, or its extension based on Eqs. (48) and (49), have general validity, nor that we have given a general explanation of the emulsion stability. The coagulation in emulsions is such a complex phenomenon, influenced by too many different factors, that according to us any attempt for formulating a *general* explanation (or criterion) is hopeless. Our treatment is theoretical and it has limitations inherent to the model used and therefore is valid only under specific conditions. It should not be applied to system where these conditions are not fulfilled. The main assumptions and limitations of the model are: the fluctuation-wave mechanism for coalescence is assumed to be operative; the surfactant transfer onto the surface is under diffusion or electro-diffusion control; parameter b defined by Eq. (37) does not account for the demicellization kinetics; etc. Only *small* perturbations in the surfactant distribution, which are due to the flow, have been considered; however, under strongly non-equilibrium conditions (like turbulent flows) we could find that new effects come into play, which may significantly alter the trend of the phenomenon.

References

Adamson, A. W., and Gast, A. P. (1997). *Physical Chemistry of Surfaces*. New York: Wiley, 6t edition.

Alexander, S. J., *Physique* 38 (1977) 983.

Arya, S.P., Air Pollution Meteorology and Dispersion, Oxford Univ. Press, New York, 1999.

Bancroft, W.D., J. Phys. Chem. 17 (1913) 514.

Barnes, H.A., J.F. Hutton and K. Walters, An Introduction to Rheology, Elsevier, Amsterdam, 1989.

Basheva, E.S., T.D. Gurkov, I.B. Ivanov, G.B. Bantchev, B. Campbell and R.P. Borwankar, Langmuir 15 (1999) 6764.

Batchelor, G.K., An Introduction of Fluid Mechanics, Cambridge Univ. Press, London, 1967.

Binks, B.P., Langmuir 9 (1993) 25.

Binks, B.P., Modern Aspects of Emulsion Science, Royal Society of Information Services, Cambridge, 1998.

Britter, R.E. and R.F. Griffiths (Editors), Dense Gas Dispersion, Elsevier Scientific Pub. Co., Amsterdam, 1982.

Chesters, A. and I.B. Bazhlekov, J. Colloid Interface Sci. 230 (2000) 229.

Chesters, A., Trans. Inst. Chem. Eng. 69 (1991) 259.

Christenson, H.K. and P.M. Claesson, Adv. Colloid Interface Sci. 91 (2001) 391.

Christenson, H.K. and P.M. Claesson, Science 239 (1988) 390.

Cristini, V., J. Blawzdziewicz and M. Loewenberg, J. Fluid Mech. 366 (1998) 259.

Danov, K.D., D.N. Petsev and N.D. Denkov, J. Chem. Phys. 99 (1993) 7179.

Danov, K.D., D.S. Valkovska and I.B. Ivanov, J. Colloid Interface Sci. 211 (1999a) 291.

Danov, K.D., I.B. Ivanov and P.A. Kralchevsky, Interfacial rheology and emulsion stability, in "Second World Congress on Emulsion", Paris, 1997, 2-2-152.

Danov, K.D., P.A. Kralchevsky and I.B. Ivanov, in "Encyclopedic Handbook of Emulsion Technology", J. Sjöblom (Editor), Marcel Dekker, New York, 2001, Ch. 26.

Danov, K.D., P.A. Kralchevsky and I.B. Ivanov, in "Handbook of Detergents. Part. A: Properties", G. Broze (Editor), Marcel Dekker, New York, 1999, pp. 303.

Davies, J. and E. Rideal, Interfacial Phenomena, Academic Press, New York, 1963.

Davies, J.T., Proceedings of the 2nd International Congress on Surface Activity, Vol. 1, Butterworths, London, 1957, p. 426.

Davis, H.T., Factors determining emulsion type: HLB and beyond, in Proc. First World Congress on Emulsion 19-22 Oct. 1993, Paris, 1993, p. 69.

Davis, R.H., J.A. Schonberg and J.M. Rallison, Phys. Fluids A1 (1989) 77.

de Gennes, P.G., Adv. Colloid Interface Sci. 27 (1987) 189.

de Gennes, P.G., C. R. Acad. Sci. (Paris) 300 (1985) 839.

de Roussel, P., D.V. Khakhar and J.M. Ottino, Chem. Eng. Sci. 56 (2001) 5511.

de Vries, A., Rec. Trav. Chim. Pays-Bas. 77 (1958) 383; Rubber Chem. Tech. 31 (1958) 1142.

Debye, P., Physik 2 (1920) 178.

Denkov, N.D., D.N. Petsev and K.D. Danov, Phys. Rev. Lett. 71 (1993) 3226.

Derjaguin, B.V., N.V. Churaev and V.M. Muller, Surface Forces, Plenum Press: Consultants Bureau, New York, 1987.

Derjaguin, B.V., Theory of Stability of Colloids and Thin Liquid Films, Plenum, Consultants Bureau, New York, 1989.

Dickinson, E., B.S. Murray and G. Stainsby, J. Chem. Soc. Faraday Trans. 84 (1988) 871.

Dukhin, S.S., G. Kretzschmar and R. Miller, Dynamics of Adsorption at Liquid Interfaces, Elsevier, Amsterdam, 1995.

Dzyaloshinskii, I.E., E.M. Lifshitz and L.P. Pitaevskii, Adv. Phys. 10 (1961) 165.

Edwards, D.A., H. Brenner and D.T. Wasan, Interfacial Transport Processes and Rheology, Butterworth-Heinemann, Boston, 1991.

Eriksson, J.C., S. Ljunggren, and P.M. Claesson, Chem. Soc. Faraday Trans. 85 (1989) 163.

Exerova, D. and P.M. Kruglyakov, Foam and Foam Films: Theory, Experiment, Application, Elsevier, New York, 1998.

Felderhof, B. U., J. Chem. Phys. 49 (1968) 44.

Griffin, J., Soc. Cosmet. Chem. 5 (1954) 4.

Gurkov, T.D. and E.S. Basheva, Hydrodynamic Behavior and Stability of Approaching Deformable Drops, in "Encyclopedia of Surface and Colloid Science", A.T. Hubbard (Editor), Marcel Dekker, New York, 2002.

Hamaker, H.C., Physics 4 (1937) 1058.

Happel, J. and H. Brenner, Low Reynolds Number Hydrodynamics with Special Applications to Particulate Media, Prentice-Hall, Englewood Cliffs, New York, 1965.

Hardy, W. and I. Bircumshaw, Proc. R. Soc. London A 108 (1925) 1.

Hetsroni, G. (Editor), Handbook of Multiphase Systems, Hemisphere Publ., New York, 1982.

Horn, R.G., O.I. Vinogradova, M.E. Mackay and N. Phan-Thien, J. Chem. Phys. 112 (2000) 6424.

Hunter, R.J., Introduction to Modern Colloid Science, Oxford Univ. Press, New York, 1993.

Israelachvili, J.N., Intermolecular and Surface Forces, Academic Press, London, 1992.

Ivanov, I.B. and D.S. Dimitrov, Colloid Polymer Sci. 252 (1974) 982.

Ivanov, I.B. and D.S. Dimitrov, in "Thin Liquid Films: Fundamentals and Applications", I.B. Ivanov (Editor), Marcel Dekker, New York, 1988, pp. 379.

Ivanov, I.B. and P. A. Kralchevsky, Colloids Surf. A 128 (1997) 155.

Ivanov, I.B., B. Radoev, E. Manev and A. Scheludko, Trans. Faraday Soc. 66 (1970) 1262.

Ivanov, I.B., B.P. Radoev, T.T. Traykov, D.S. Dimitrov, E.D. Manev and C.S. Vassilieff, in "Proceedings of the International Conference on Colloid Surface Science", E. Wolfram (Editor), Akademia Kiado, Budapest, 1975, p. 583.

Ivanov, I.B., K.D. Danov and P.A. Kralchevsky, Colloids and Surfaces A 152 (1999) 161.

Ivanov, I.B., Pure Appl. Chem. 52 (1980) 1241.

Keesom, W.H., Proc. Amst. 15 (1913) 850.

Kim, S. and S.J. Karrila, Microhydrodynamics: Principles and Selected Applications, Butterworth-Heinemann, Boston, 1991.

Klitzing, R. and H.J. Müller, Curr. Opinion in Colloid & Surface Sci. 7 (2002) 42.

Kralchevsky, P.A. and N.D. Denkov, Chem. Phys. Lett. 240 (1995) 385.

Kralchevsky, P.A., K.D. Danov and N.D. Denkov, in "Handbook of Surface and Colloid Chemistry", K.S. Birdi (Editor), CRC Press, New York, Second Edition, 2002, Ch. 11.

Kralchevsky, P.A., K.D. Danov, G. Broze and A. Mehreteab, Langmuir 15 (1999) 2351.

Landau, L.D. and E.M. Lifshitz, Electrodynamics of Continuous Media, Pergamon Press, Oxford, 1960.

Landau, L.D. and E.M. Lifshitz, Fluid Mechanics, Pergamon Press, Oxford, 1984.

Langmuir, I., J. Chem. Phys. 6 (1938) 873.

Leal, G.L., Laminar Flow and Convective Transport Processes. Scaling Principles and Asymptotic Analysis, Butterworth-Heinemann, Boston, 1992.

Lifshitz, E.M., Soviet Phys. JETP (Engl. Transl.) 2 (1956) 73.

London, F., Z. Physics 63 (1930) 245.

Maldarelli, Ch. and R. Jain, in "Thin Liquid Films: Fundamentals and Applications", I.B. Ivanov (Editor), Marcel Dekker, New York, 1988, pp. 497.

Manev, E.D., S. V. Sazdanova, C.S. Vassilieff and I.B Ivanov, Ann. Univ. Sofia Fac. Chem. 71(2) (1976/1977) 5.

Manev, E.D., S.V Sazdanova and D.T. Wasan, J. Colloid Interface Sci. 97 (1984) 591.

Pashley, R.M., P.M. McGuiggan, B.W. Ninham and D.F. Evans, Science 229 (1985) 1088.

Paunov, V.N., S.I. Sandler and E.W. Kaler, Langmuir 17 (2001) 4126.

Prud'homme, R.K. and S.A. Khan (Editors), Foams: Theory, Measurements and Applications, Marcel Dekker, New York, 1996.

Radoev, B.P., D.S. Dimitrov and I.B. Ivanov, Colloid Polym. Sci. 252 (1974) 50.

Reynolds, O., Phil. Trans. Roy. Soc. (Lond.) A177 (1886) 157.

Richetti, P. and P. Kékicheff, Phys. Rev. Lett. 68 (1992) 1951.

Russel, W.B., D.A. Saville and W.R. Schowalter, Colloidal Dispersions, Univ. Press, Cambridge, 1989.

Sche, S. and H.M. Fijnaut, Surface Sci. 76 (1976) 186.

Scheludko, A. and D. Exerowa, Kolloid-Z. 165 (1959) 148.

Scheludko, A., Proc. K. Akad. Wetensch. B 65 (1962) 87.

Schmitz, K.S., Langmuir 12 (1996) 3828.

Schramm, L.L. (Editor), Suspensions: Fundamentals and Applications in the Petroleum Industry, American Chemical Society, Washington, 1996.

Shinoda, K. and S. Friberg, Emulsion and Solubilization, Wiley, New York, 1986.

Singh, G., G.J. Hirasaki and C.A. Miller, J. Colloid Interface Sci. 184 (1996) 92.

Sjoblom, J. (Editor), Emulsions and Emulsion Stability, M. Dekker, New York, 1996.

Slattery, J.C., Interfacial Transport Phenomena, Springer-Verlag, New York, 1990.

Slattery, J.C., Momentum, Energy, and Mass Transfer in Continua, R.E. Krieger, Huntington, New York, 1978.

Stoyanov, S.D. and N.D. Denkov, Langmuir, 17 (2001) 1150.

Tambe, D.E. and M.M. Sharma, J. Colloid Interface Sci., 147 (1991) 137.

Tambe, D.E. and M.M. Sharma, J. Colloid Interface Sci., 157 (1993) 244.

Tambe, D.E. and M.M. Sharma, J. Colloid Interface Sci., 162 (1994) 1.

Traykov, T.T and I.B. Ivanov, Int. J. Multiphase Flow 3 (1977) 471.

Traykov, T.T., E.D. Manev and I.B. Ivanov, Int. J. Multiphase Flow 3 (1977a) 485.

Valkovska D.S., K.D. Danov and I.B. Ivanov, Adv. Colloid and Interface Sci. 96 (2002) 101.

Valkovska, D.S. and K.D. Danov, J. Colloid Interface Sci., 223 (2000) 314.

Valkovska, D.S. and K.D. Danov, J. Colloid Interface Sci., 241 (2001) 400.

Valkovska, D.S., K.D. Danov and I.B. Ivanov, Colloids and Surfaces A 156 (1999) 547.

Verwey, E.J.W. and J.Th.G. Overbeek, Theory and Stability of Lyophobic Colloids, Elsevier, Amsterdam, 1948.

Vrij, A., Disc. Faraday Soc. 42 (1966) 23.

Yiantsios S.G. and R.H. Davis, J. Colloid Interface Sci. 144 (1991) 412.

Yushchenko, V. S., V.V. Yaminsky and E.D. Shchukin, Colloid Interface Sci. 96 (1983) 307.

Surfactant Effects on Mass Transfer in Liquid-Liquid Systems

M. Alcina Mendes*

Department of Chemical Engineering and Chemical Technology,
Imperial College London, SW7 2AZ, UK

Abstract. This paper reviews the work done by the author and co-workers on the effect of surfactants on mass transfer in binary and ternary liquid-liquid systems. A Schlieren optical apparatus has been used to visualise the selected organic-aqueous interfaces during the mass transfer process, when the aqueous phase was "clean" or "contaminated" by soluble ionic and non-ionic surfactants. Molar fluxes, in the same liquid-liquid systems, have been measured with a Mach-Zehnder interferometer. Results obtained in the laboratory and under microgravity conditions are included. The most significant finding is that the addition of some surfactants to the partially miscible binary liquid-liquid systems investigated can induce or increase interfacial convection which enhances the initial mass transfer rates in comparison with values predicted by Fick's law. This effect is of great relevance to industrial processes where surfactants may be used advantageously to manipulate interfacial stability and particularly in space applications, where it may be exploited to increase mass transfer in compensation for the lack of gravitational convection effects.

1   Introduction

Mass transfer across interfaces is ubiquitous in industrial processes. For instance, it occurs in separation processes used in the field of biotechnology, in the extraction of metals from aqueous solutions, in the chemical and pharmaceutical industries and also in the treatment of effluents from the same plants. However, in the design of the contacting equipment the role of the interface and of interfacial phenomena are not taken into account. Take the example of the industrial effluent streams. These are usually not made up of pure components but are "soups" containing different chemicals and chemicals with surfactant properties. However, in the design of liquid-liquid contactors those streams are considered to be "clean" and interfacial phenomena (such as Marangoni and gravitational convection), which may exist, are not taken into account.

Due to the ever increasing demands for improvement in the efficiency of industrial processes fundamental research needs to be carried out to obtain a clear understanding of the role of the interface and its inherent phenomena, on the mass transfer process.

For a fixed interfacial area, mass transfer rates may be increased if Marangoni convection is present, as has been well acknowledged for many years (see e.g. Sawistowski, 1971, Berg, 1972 and Perez de Ortiz, 1991). The interaction between gravitational and Marangoni effects in ternary systems has also been found to be beneficial to the mass transfer process (e.g. Berg & Morig, 1969).

On the other hand, the presence of surfactants in systems has been usually accepted as being disadvantageous to mass transfer (e.g. Blokker, 1957, Davies and Wiggill, 1960, Mudge and Heideger, 1970, Komasawa et al, 1972, Godfrey and Slater, 1995, Brodkorb and Slater, 2001) as their adsorption at the interface may create a barrier to the transfer of other species or damp any interfacial convection present. In contrast, research, which concludes that the adsorption/desorption of surfactants at an interface may produce Marangoni convection, which enhances mass transfer, has also been published (e.g. Nakache et al., 1983, Nakache and Raharimalala, 1988, Aunins et al., 1993, Bennett et al., 1996, Agble and Mendes-Tatsis, 2000). The process of adsorption and desorption of a surfactant at a liquid-liquid interface changes the interfacial properties and, depending on the properties of the system, Marangoni convection may be initiated. This convection will aid surface renewal and enhance mass transfer.

Therefore, for the optimisation of the design of liquid-liquid contacting equipment and the improvement of mass transfer processes the interfacial hydrodynamics needs to be well understood, so that it can be taken into account at a design stage and if necessary (and possible!), enable the manipulation of the interfacial behaviour to enhance the mass transfer rates.

With that in mind the author and co-workers have been carrying out experiments with many liquid-liquid systems and in this paper the results obtained in recent years are summarised. Sections 2 and 3 include theory and the definition of interfacial convection, which give the background for the analysis of the experimental results shown later. Section 4 describes the experimental equipment and methods used to obtain the results presented in: Section 5, for liquid-liquid partially miscible binary systems with and without surfactants; Section 6, for the same systems under microgravity conditions; and Section 7, for ternary systems with and without surfactants. Section 8 covers work previously done on stability criteria to predict Marangoni convection in liquid-liquid systems. In Section 9 some of the results presented in the previous sections are discussed. Relevant results obtained in the field of Biotechnology by the author and co-workers are mentioned in Section 10. Conclusions and future work follow in Section 11.

* The author acknowledges the contribution made by the co-workers and the financial support of the EPSRC.

2   Diffusional Mass Transfer

When mass transfer occurs by diffusion only, the diffusional molar flux can be predicted using Fick's law (1855)

$$N_A = -D_A \frac{dC_A}{dx} \qquad (1)$$

where D_A is the diffusivity of species A in the medium under consideration and dC_A/dx is the concentration gradient.

The application of Fick's law (Equation 1) to unsteady diffusional mass transfer leads to Fick's second law

$$\frac{\partial C_A}{\partial t} = -D_A \frac{\partial^2 C_A}{\partial x^2} \qquad (2)$$

where t indicates time, and which can be solved for a semi-infinite case with appropriate boundary conditions (Crank, 1975) to give

$$C_A = C_{A_i} \, erfc \left(\frac{x}{2 \sqrt{D_A \, t}} \right) \qquad (3)$$

This equation gives a prediction of the variation of C_A with time t at a given distance x away from the interface.

When diffusional mass transfer is accompanied by interfacial convection the predictions given by the above equations are no longer correct. An understanding of the causes for interfacial convection is given in the next section.

3   Interfacial convection

The occurrence of destabilising concentration and/or temperature gradients at a liquid-liquid interface while a mass transfer process is taking place causes local variations in interfacial tension, which may lead to Marangoni convection (Thomson, 1855; Marangoni, 1865, 1871). In addition, temperature gradients may also produce density gradients and cause gravitational convection.

Lewis & Pratt, in 1953, were the first to report that the observed Marangoni convection in their experimental ternary systems was beneficial to liquid-liquid extraction processes because it increased mass transfer rates. The effect of density gradients on interfacial convection was studied by several researchers including Berg & Morig (1969), who investigated the interaction between buoyancy and interfacial tension driven effects in ternary systems. The combined interfacial convection was also seen to be beneficial to mass transfer processes.

There are processes, however, which do not benefit from interfacial convection such as the formation of crystals. In this case interfacial convection causes the development of unwanted striations in the crystals produced and hence has a detrimental effect on the process.

It is, therefore, important to understand interfacial phenomena well to enable the development of means to manipulate the interfacial conditions in order to enhance or suppress interfacial convection.

One type of substance that affects interfacial conditions is a surfactant. The presence of surfactants in liquid-liquid systems has traditionally been viewed as detrimental to mass transfer because it has been found they either damp interfacial convection or cause a barrier effect to mass transfer. On the other hand, mass transfer may also be increased because of the presence of surfactants, as it will be shown in the Sections 5 and 6.

4   Experimental Equipment and Methods

4.1 Schlieren/Mach Zehnder Interferometer

Figure 1 shows a diagram of the combined Schlieren/Mach-Zehnder interferometer apparatus, used to obtain the experimental results that are to be presented in the next sections. This apparatus can be changed, by the insertion of mirrors (14) and (15), from the interferometer configuration to become a Schlieren optical system to enable it to be used for the visualisation of the interfaces across which mass transfer occurs. A detailed description of the integrated apparatus and techniques may be found in Agble and Mendes-Tatsis (2000).

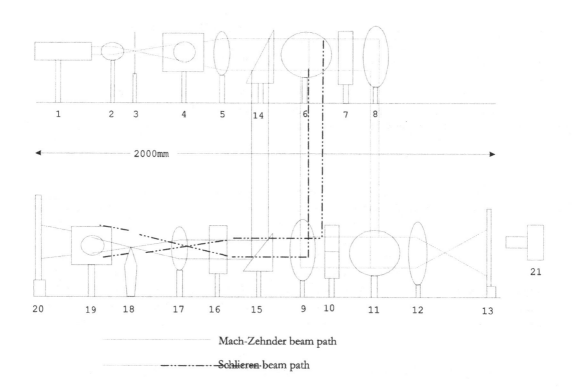

Figure 1 - Integrated Schlieren/Mach-Zehnder optical apparatus (Agble and Mendes-Tatsis, 2000).

Labels:

1	He-Ne Laser	2	Microscope Objective
3	25µ Pinhole	4,19	Circular Diaphragms
5	f500 mm Plano-Convex Lens	6,11	Semi-Reflecting Mirrors
7	Interferometer Reference Cell	8,9,14,15	Fully Reflecting Mirrors
10	Interferometer Test Cell	12	f125 mm Plano-Convex Lens
13,20	Translucent Screens	16	Schlieren Drop Cell
17	F75 mm Bi-Convex Lens	18	Tapered Knife Edge
21	Photographic/Video Camera		

Interferometry method. This method involves the analysis of a fringe pattern which changes with time reflecting the corresponding variations of the refractive index of the medium caused by the mass transfer process (Caldwell, 1957). This enables the determination of the concentration of the species transferred at any location close to the interface (up to 250μm) at various times.

From the experimentally measured concentrations, $C_A(t, x)$, values for the experimental molar fluxes as a function of time may be obtained. These results can then be compared with the ones predicted by Fick's theory (Equation 3).

In the next sections experimental results obtained with the Schlieren and Mach Zehnder methods are presented. It becomes apparent that the Schlieren visualisation of interfaces is an excellent method to give a clear idea of the interfacial behaviour during mass transfer, from which one can infer its effect on mass transfer.

5   Binary Systems with and without Surfactants

Mass transfer in quiescent, binary partially miscible organic-aqueous systems has been investigated when the aqueous phase was pure water and when it contained a surfactant (Agble, 1998). In all cases, mass transfer was always from the organic to the aqueous phase and the surfactant was always added, in concentrations below the CMC, to the aqueous phase, but was also soluble in the organic phase.

From the experimental results obtained three liquid-liquid systems have been selected to be included here: aniline/water, ethylacetate/water and ethylacetoacetate/water. The selected surfactants are: SDS (sodium dodecyl sulphate), which is anionic, DTAB (dodecyl tri-methyl ammonium bromide), which is cationic, and Atlas G1300 (polyoxyethylene triglyceride ester) which is non-ionic.

The Schlieren images in Figure 2 show drops of each of the organic phases investigated immersed in either pure water or in an aqueous phase "contaminated" with a surfactant. These photos were taken 10s after the formation of the drop. At the aniline/water interface no instabilities are observed (case 1a) while the presence of instabilities at the ethyl acetate/water (case 2a) and ethylacetoacetate/water (case 3a) interfaces are very obvious.

In Figure 3, a comparison of the interferometrically measured fluxes with the theoretical ones shows that the transfer of aniline into pure water follows Fick's diffusional theory. This is expected since the interface was shown to be stable in the Schlieren image (case 1a) in Figure 2, and therefore mass transfer in this case is by diffusion only.

If, however, SDS or DTAB is added to the aqueous phase Marangoni convection is seen to have been induced in the case the aniline/water interface, as can be seen in Figure 2, cases 1b,c. In the other two systems (cases 2b,c and 3b,c), the addition of SDS or DTAB produces an increase in the convection intensity in comparison to the one observed when the water was "clean". On the other hand, the addition of ATLAS G1300 has either no effect (case 1d) on the interfacial stability or damps (cases 2d and 3d) any convection present in the "clean" system.

A comparison between the predicted and the various interferometric results obtained for the molar fluxes under different aqueous phase conditions is shown in Figure 4 for the case of the transfer of ethylacetate into water. The transfer into pure water is higher than theoretically predicted, as expected, since the interfacial convection seen in Figure 2 (case 2a), increases the mass transfer occurring across the interface. The addition of ATLAS G1300 shows that it hardly has any effect on the experimental fluxes when compared with the "clean" system, despite the suppression of the interfacial convection shown in Figure 2 (case 2d). The most interesting results, though, are the ones for the cases of DTAB and SDS, which produce much higher initial fluxes than for the case of the "clean" system.

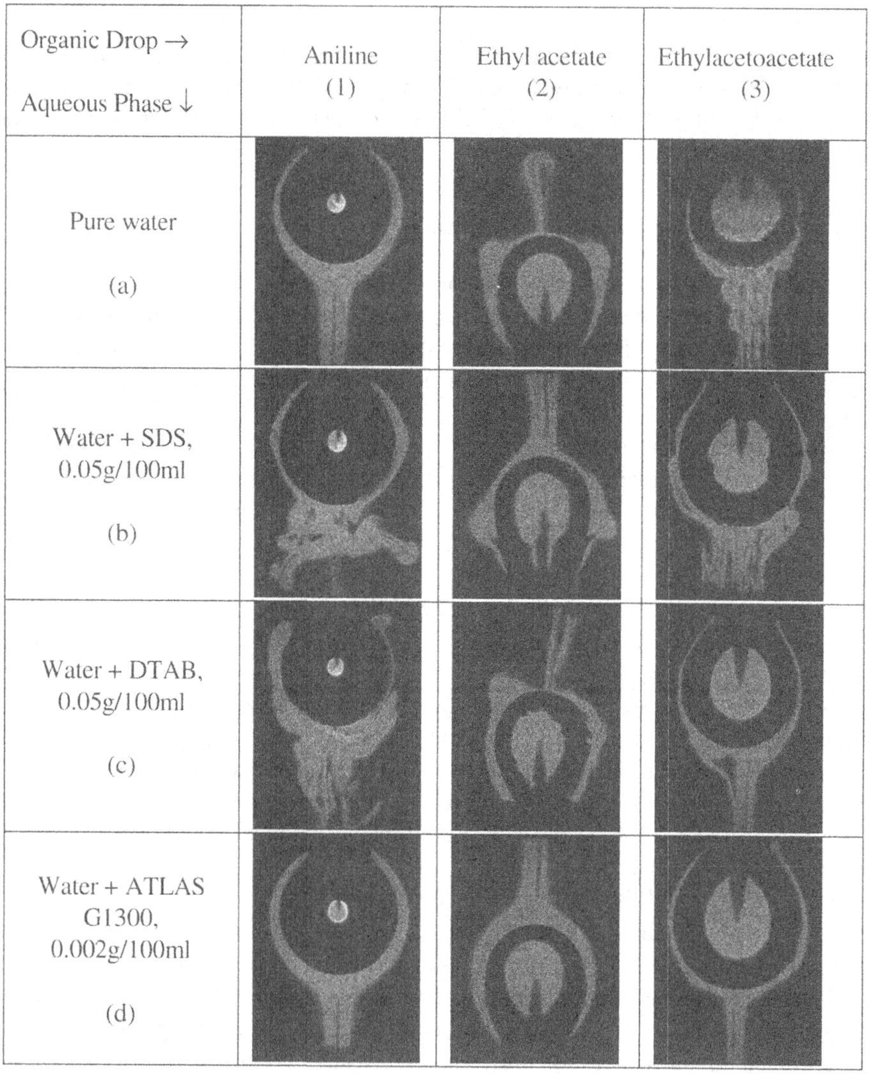

Organic Drop → Aqueous Phase ↓	Aniline (1)	Ethyl acetate (2)	Ethylacetoacetate (3)
Pure water (a)			
Water + SDS, 0.05g/100ml (b)			
Water + DTAB, 0.05g/100ml (c)			
Water + ATLAS G1300, 0.002g/100ml (d)			

Figure 2 – Schlieren images of different systems when mass transfer is from the organic drops to the outer aqueous phase, when "clean" or when contaminated by different surfactants (Agble and Mendes-Tatsis, 2000).

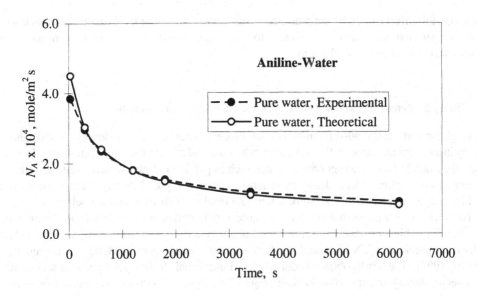

Figure 3 – Comparison between the predicted and theoretical molar fluxes vs. time for the transfer of aniline into water (Agble and Mendes-Tatsis, 2000).

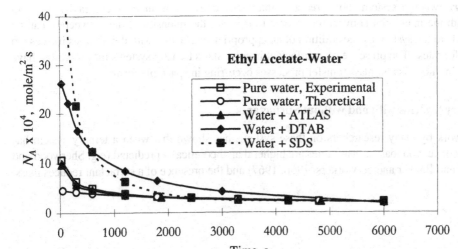

Figure 4 - Molar flux vs. time for the transfer of ethyl acetate into different aqueous phases (Agble and Mendes-Tatsis, 2000).

Therefore, this effect caused by the presence of surfactants has strong implications on mass transfer processes: selected surfactants may be added to a particular system to induce Marangoni convection in order that mass transfer is enhanced

6   Binary Systems with and without Surfactants under Microgravity

Although the surfactants' adsorption/desorption process is not affected by density the existence of gravity in a system may modify the interfacial behaviour and affect interfacial convection. In recent years, the availability of microgravity facilities such as parabolic flights to investigate the effect of Marangoni convection, independently from density effects (i.e. pure Marangoni convection) on the fluid hydrodynamics has produced some interesting results which are presented below.

To check whether gravitational effects are necessary to initiate the interfacial convection during mass transfer in binary partially miscible liquid-liquid systems, experiments have been carried out, under microgravity (ESA/NASA parabolic flights), with "clean" systems (Mendes-Tatsis and Perez de Ortiz, 1992). The results obtained confirmed that interfacial tension gradients were sufficient for the onset of Marangoni convection in binary liquid-liquid interfaces undergoing mass transfer in the absence of gravity.

Experiments with a surfactant added to the aqueous phase have also been carried out under the microgravity (ESA/NASA parabolic flights) (Agble and Mendes-Tatsis, 1997 and Mendes-Tatsis, 2000). The results were very interesting and can be seen in Figure 5 for the case of ethylacetoace-tate/water.

This system is known to be gravitationally and Marangoni unstable, and the strong interfacial convection observed on Earth is shown in Figure 5a for the simultaneous transfer of ethylaceto-acetate into water and water into ethylacetoacetate. Under microgravity conditions (Figure 5b) the intensity of the interfacial convection observed is reduced (only Marangoni convection exists) when compared to that observed on Earth, where gravitational convection is also present. However, the addition of SDS to the aqueous phase under microgravity (Figure 5c) shows interfacial convection, which is of similar (or even stronger) intensity to that observed on Earth.

In summary: when a system shows reduced interfacial convection under microgravity, in comparison with the interfacial convection observed on Earth, the intensity of the convection can be "re-instated" to the system by the addition of an appropriate surfactant with the obvious increase in mass transfer rates. The presence of a surfactant in some mass transfer systems may therefore have significant implications on mass transfer processes occurring in space platforms.

7   Ternary Systems with and without Surfactants

Previous work by many researchers on these systems has shown that when a ternary system has interfacial convection mass transfer rates are higher than theoretically predicted (e.g. Sherwood and Wei, 1957 and Bakker and co-workers, 1966, 1967) and the presence of a surfactant reduces mass

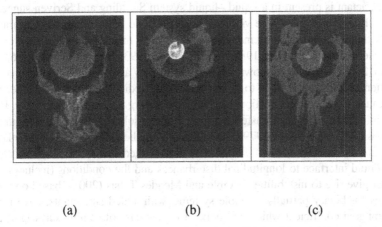

(a) (b) (c)

Figure 5 - Simultaneous transfer of ethylacetoacetate into water and water into ethylacetoacetate: (a) under 1-g conditions; (b) under microgravity conditions when the water was "clean" and (c) under microgravity conditions when the water contained SDS (Mendes-Tatsis, 2000).

transfer rates below the predicted ones (e.g. Blokker, 1957, Davies and Wiggill, 1960, Mudge and Heideger, 1970, Komasawa et al, 1972, Godfrey and Slater, 1995, Brodkorb and Slater, 2001). Recent work by the author and co-workers has found similar results, but in some cases, the presence of surfactants reduced mass transfer rates when compared to the clean system, but they were still higher than the ones predicted by diffusional theory (Pursell and Mendes, 2003). These results were also confirmed visually by the Schlieren images, which show interfacial convection, but of a lower intensity than in the clean case.

In summary: The beneficial surfactant effects on mass transfer observed with the simpler binary systems may be in competition with the multicomponent effects present in ternary systems. It is obvious that not enough is known about surfactant effects on mass transfer systems and more research needs to be carried out.

It is also clear that if benefits are to be obtained from the understanding of those surfactant effects, it is important to be able to predict the occurrence of interfacial convection in the absence and presence of surfactants. The work done in this direction is mentioned in the next section.

8 Stability criteria

Sternling and Scriven (1959) were the first to explain and produce stability criteria to predict interfacial stability for the case of a transferring solute between two immiscible phases i.e. ternary liquid-liquid systems. Their criteria have been confirmed many times (e.g. Bakker et al., 1966 and Orell and Westwater,1962) but there have also been some cases where the predictions did not agree with experimental observations (e.g. Linde and Schwarz, 1964 and Maroudas and Sawistowski, 1964).

For binary systems Perez de Ortiz and Sawistowski (1973a) proposed stability criteria, which were obtained following a similar mathematical treatment to the one by Sternling and Scriven and applied them successfully to several binary systems Perez de Ortiz and Sawistowski (1973b). However, they also found some disagreements with experimental results.

When a surfactant is present in a liquid –liquid system Sternling and Scriven suggested that it would have such an effect on surface viscosity as to preclude the occurrence of Marangoni convection. While this was confirmed by many researchers there have also been several experimental results (Nakache and Raharimalala, 1988, Aunins et al., 1993, Bennett et al., 1996, Agble and Mendes-Tatsis, 2000), which have shown that surfactants may also initiate interfacial convection.

Several attempts have been made to obtain criteria to predict the conditions under which surfactants may cause Marangoni convection to occur. Hennenberg and co-workers (1979, 1980, 1981) developed stability criteria for the transfer of a surface active solute across a liquid-liquid interface; Gouda and Joos (1981), Sörensen (1979), Sanfeld and Steinchen (1984), Chu and Velarde (1989) and Nakache and co-workers (1983, 1988, 1991) also studied mathematically the stability of a planar liquid-liquid interface to longitudinal disturbances and the conditions (including surfactant properties) that give rise to instabilities. Agble and Mendes-Tatsis (2001), based on their experimental findings for binary partially miscible systems, with added surfactants, have proposed an empirical Marangoni coefficient which was defined for the case of soluble surfactants and which represents a ratio of the interfacial tension effects (which produce interfacial motion), to the viscous and related effects (that limit surface movements). Comparisons between several stability criteria are included in the same reference.

Unfortunately, despite the efforts by all the researchers mentioned above and others, interfacial convection in liquid-liquid systems cannot yet be accurately predicted. More work is in progress by the author and co-workers to achieve better predictions for the case of binary systems (Slavchev and Mendes, 2003) and binary systems with surfactants (Slavchev, Kalitzova-Kurteva and Mendes, 2003).

9   Discussion

The presence of surfactants in the mass transfer systems presented has been shown to affect the mass transfer process by inducing, reducing or suppressing interfacial convection. Since interfacial convection is caused by interfacial tension variations and the addition of a surfactant to a system alters interfacial tension, an examination of the effect surfactants have on the interfacial tension of the systems is of interest.

During the mass transfer processes described above while the surfactants adsorb and transfer across the liquid-liquid interface, they cause variations in the interfacial tension. If surfactant diffusion, adsorption and desorption rates are neglected, then the surfactant interfacial concentration is only a function of its bulk concentration which decreases as the surfactant transfers across the interface. Therefore, from the analysis of the plots of equilibrium interfacial tension vs. the bulk surfactant concentration, an understanding of the occurrence of Marangoni convection may be possible.

Equilibrium interfacial tension results are shown in Figures 6, 7 and 8 for aniline/water, ethyl acetate/water and ethyl acetoacetate-water systems, using three different surfactants, SDS, DTAB and Atlas G1300 at various concentrations. In these figures each interfacial tension value is an average of five measured values, from experiments carried out at 25C.

The ionic surfactants SDS and DTAB were seen to initiate Marangoni convection, when the pure system was stable, or increase its intensity if the uncontaminated system was already unstable (see Figure 2). Analysis of the curves of interfacial tension vs. surfactant concentration (Agble and Mendes-Tatsis, 2001) shows that generally the ionic surfactants and the non-ionic surfactants investigated produce large changes in the interfacial tension but that in the case of ATLAS G1300 there is a slight difference: while the ionic surfactants show a high gradient over a wide concentration range, ATLAS G1300 shows a very high initial gradient which rapidly levels out (except in the case of aniline/water). This may be interpreted that during mass transfer, the presence

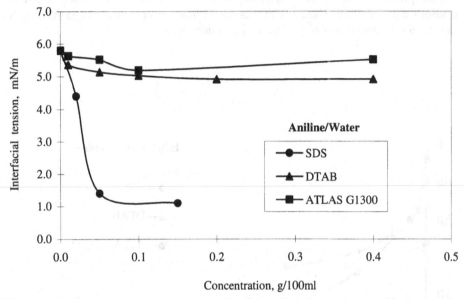

Figure 6 Interfacial tension vs. concentration of surfactant in the water, for the aniline/water system (Agble and Mendes-Tatsis, 2001)

of ionic surfactants result in a sustained change in interfacial tension over the mass transfer process, which will promote Marangoni effects, while the non-ionic surfactant gives a large initial change that is not sustained over the mass transfer process and therefore is unlikely to produce Marangoni convection.

One of the reasons for this difference in behaviour may be the size and structure of the surfactant molecules. Longer chain surfactants are generally more effective at reducing the interfacial tension, because of their greater interaction with the phase molecules at the interface. Therefore, a very small amount of surfactant causes a large decrease in the interfacial tension, and a further increase in the surfactant concentration will produce only small changes in the interfacial tension. This does not provide the appropriate conditions for the occurrence of Marangoni convection. This is in contrast with the situation for the ionic surfactants illustrated for example in Figure 7, where there is a sustained slope of the SDS and DTAB curves which will promote suitable conditions for the onset of Marangoni convection. Similar findings have been reported by Lyford et al. (1992) for the case of different alcohols. The reason for the damping of interfacial convection by ATLAS G 1300 might also be to do with being non-ionic (Van Voorst Vader, 1960) and with its large size, forming a monolayer which makes the interface more rigid and reduces interfacial motion.

Unlike the other cases, in the aniline/water system the addition of Atlas G1300 does not have much effect on the interfacial tension, indicating the unlikelihood of the system having an unstable interface, as shown in the Schlieren images (Figure 2, case 1c). However, even the small changes in interfacial tension values shown for SDS and DTAB are sufficient for the onset of interfacial tension in comparison with the "clean" aniline/water system.

Therefore the addition of a surfactant to a liquid-liquid system may either promote Marangoni convection or dampen interfacial motions, depending on the size of the surfactant molecule and its tendency to adsorb at the interface. These two effects may compete with each other and the overall stability of a system will depend on which of these effects dominates.

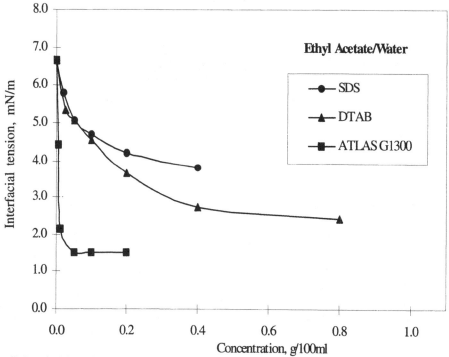

Figure 7 Interfacial tension vs. concentration of surfactant in the water, for the ethyl acetate/water system (Agble and Mendes-Tatsis, 2001)

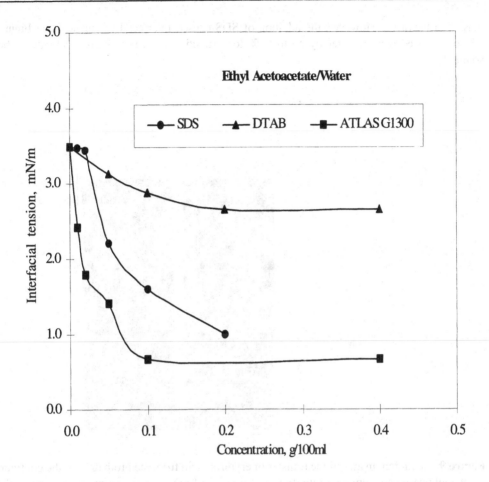

Figure 8 Interfacial tension vs. concentration of surfactant in the water, for the ethyl acetoacetate/water system (Agble and Mendes-Tatsis, 2001)

Further work needs to be carried out to improve the simplified analysis presented by investigating, for example, the effect of dynamic interfacial tension on the interfacial convection.

10 Biotechnological industries

The cases mentioned above are typical of simple liquid-liquid extraction systems, which can be found in the traditional chemical industries. However, extraction systems in the biotechnological industries are more complicated because of the presence of microbial cells and surface active compounds produced during fermentation. The adsorption to the interface of soluble and insoluble surface-active compounds present in the fermentation broth usually causes a mass transfer resistance that reduces the transfer rates (Pursell et al., 2003a, 2003b, 2003c). This is in agreement with what has been found (Pursell et al., 1999) for the case of the extraction of the antibiotic erythromycin-A from a buffer solution and from filtered fermentation broth into 1-decanol: the presence of the naturally occurred soluble surface active compounds in the filtered broth led to a 45% reduction in mass transfer rates. However, the addition of SDS caused interfacial turbulence (see Figure 9) which enhanced mass transfer by up to 55% for extraction from both buffer and filtered broth solutions.

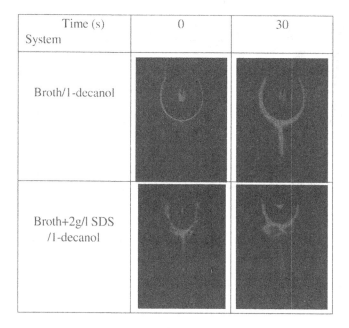

Figure 9 Schlieren images of the transfer of erythromycin from the broth drop to the continuous 1-decanol phase and from the broth drop containing 2g/l SDS to the continuous 1-decanol phase (Pursell et al., 1999)

Increases in mass transfer rates have been obtained before (Lye and Stuckey, 2001) with similar systems but interfacial convection had not been observed and the explanation put forward was that the increase in rates was caused by an interaction/aggregation between the surfactant and the solute at the interface.

It is obvious that more effort needs to be put into understanding complex multicomponent systems where some of the components have surface-active properties.

11 Conclusions and Future Work

The most important achievement of this research has been the finding that the addition of some surfactants to the binary liquid-liquid systems investigated can induce or increase interfacial convection that enhances greatly the initial mass transfer rates in comparison with values predicted by Fick's law. This shows that surfactants can be used as a means to "manipulate" the interfacial region to obtain variations in interfacial parameters, which induce interfacial convection and produce an enhancement in mass transfer. This surfactant effect is of particular importance under microgravity because it may be exploited to increase mass transfer in compensation for the lack of gravitational convection effects, as was shown with the ethylacetoacetate/water system. These findings need to be explored further, because they are of great importance to mass transfer processes on Earth and also for Space applications.

The effect surfactants have on ternary liquid-liquid systems' unstable interfaces is of either reducing or suppressing completely any interfacial convection present. In turn, this causes the mass transfer rates to be reduced.

Therefore the addition of a surfactant to a liquid-liquid system may either promote Marangoni convection or dampen interfacial motions, depending on the type of liquid-liquid system and surfactant. Stability criteria have been established for many years for ternary and binary systems and when the solute has surface-active properties, but they are still a long way from being able to predict the interfacial behaviour accurately. Plots of the variation of equilibrium interfacial tension with surfactant concentration may be used to give a good indication of the interfacial behaviour during mass transfer.

The author and co-workers have recently started work on the effect of the dynamic interfacial tension on the interfacial behaviour in systems undergoing mass transfer and this may bring more light to the topic presented here.

In conclusion, more research needs to be carried out to understand the effect of surfactants on interfacial behaviour, which leads to increases in mass transfer rates - to be able to specify a surfactant that will enhance mass transfer in a particular process is an appealing prospect.

References

Agble, D. (1998). Interfacial Mass Transfer in Binary Liquid-Liquid Systems with Added Surfactants. *PhD Thesis*, Imperial College, University of London.

Agble, D., and Mendes-Tatsis, M. A. (1997). Surfactant Induced Marangoni Convection under Microgravity. In *Proceedings of Joint Xth European and Vith Russian Symposium on Physical Sciences in Microgravity* (St Petersburg, Russia), 108-117.

Agble, D., and Mendes-Tatsis, M. A. (2000). The effect of surfactants on interfacial mass transfer in binary liquid-liquid systems. *Int. J. Heat & Mass Transfer*, 43: 1025-1034.

Agble, D., and Mendes-Tatsis, M. A. (2001). The prediction of Marangoni convection in binary liquid-liquid systems with added surfactants. *Int. J. Heat & Mass Transfer*, 44: 1439-1449.

Aunins A H, Browne E P, Hatton T A. (1993). Interfacial Transport Resistances at Surfactant-Laden Interfaces, In *Proceedings of the International Solvent Extraction Conference 1993*, York, United Kingdom, 1704-1711.

Bakker C. A. P., van Buytenen P. M., and Beek W. J. (1966). Interfacial Phenomena and Mass Transfer, *Chemical Engineering Science* 21: 1039-1046.

Bakker C. A. P., Fentener van Vlissingen F. H., and Beek W. J. (1967).The Influence of the Driving Force in Liquid-Liquid Extraction - A Study of Mass Transfer With and Without Interfacial Turbulence Under Well Defined Conditions. *Chemical Engineering Science* 22: 1349-1355.

Bennett D. E., Gallardo B. S, and Abbott N. L. (1996). Dispensing Surfactants from Electrodes: Marangoni Phenomenon at the Surface of Aqueous Solutions of (11 Ferrocenylundecyl) trimethyl ammonium Bromide. *Journal of the American Chemical Society* 118: 6499-6505.

Berg, J.C. (1972). Interfacial Phenomena in Fluid Phase Separation Processes. In *Recent Developments in Separation Science*, 2, CRC Press, 1-29

Berg, J. C., and Morig, C. R. (1969). Density Effects in Interfacial Convection. *Chemical Engineering Science*, 24: 937-945.

Blokker, P.C. (1957). On mass transfer across liquid/liquid interfaces in systems with and with out surface active agents. *Proc. 2nd International Congress of Surface Activity*, 1: 503-510.

Brodkorb, M. J. and Slater, M. J. (2001). Multicomponent and Contamination Effects on Mass Transfer in a Liquid-Liquid Extraction Rotating Disc Contactor .*Trans IChemE*, 79 A: 335-346.

Caldwell, C. S., Hall J. R., and Babb A. L. (1957). Mach-Zehnder interferometer for diffusion measurements in volatile liquid. *Review of Scientific Instruments* 28: 816-827.

Chu X. L., Velarde M. G. (1989). Transverse and Longitudinal Waves Induced and Sustained by Surfactant Gradients at Liquid-Liquid Interfaces. *Journal of Colloid and Interface Science* 131(2): 471-484

Crank, J. (1975). *The Mathematics of Diffusion*, Oxford University Press, London.

Davies, J. T., Wiggill, J. B. (1960). Diffusion Across the Oil/Water Interface. In *Proceedings of the Royal Society (London)*, A255: 277- 291.

Fick, A.E, (1855). On Liquid Diffusion, *Philosophical Magazine* 10: 30-39.

Godfrey, J.C. and Slater, M.J. (1994). *Liquid-Liquid Extraction Equipment*, John Wiley & Sons.

Gouda J.H., and Joos P. (1975). Application of Longitudinal Wave Theory to Describe Interfacial Instability. *Chemical Engineering Science* 30: 521-528.

Hennenberg M., Bisch P. M., Vignes-Adler M., Sanfeld, A. (1979). Mass Transfer, Marangoni and Instability of Interfacial Longitudinal Waves I. Diffusional Exchanges. *Journal of Colloid and Interface Science* 69 (1): 128-137

Hennenberg M., Bisch P. M., Vignes-Adler M., Sanfeld A. (1980). Mass Transfer, Marangoni and Instability of Interfacial Longitudinal Waves I. Diffusional Exchanges and Adsorption-Desorption Processes. *Journal of Colloid and Interface Science* 74 (2), 495-508.

Hennenberg M., Sanfeld A., Bisch P. M. (1981). Adsorption-Desorption Barrier, Diffusional Exchanges and Surface Instabilities of Longitudinal Waves for Aperiodic Regimes. *Journal of the American Institute of Chemical Engineers* 27(6): 1002-100.

Komasawa, I., Saito, T., and Otake, T. (1972). Mass transfer across the liquid-liquid interface. Interfacial turbulence and its elimination. *Inst. Chem. Eng.*, 12, 2, 345-351.

Lewis, J. B., and Pratt, H. R. C. (1953). Oscillating Droplets, *Nature*, 171: 1155-1156.

Linde H. and Schwarz E. (1964). Uber groBraumige Rollzellen der freien Grenzflachenkonevktion. *Monatsberitche Deutsche Akademie der Wissenschaften zu Berlin* 7: 330-338.

Lye G.J., and Stuckey D. C. (2001). Extraction of Erythromycin-A Using Colloidal Liquid Aphrons: Part II. Mass Transfer Kinetics. *Chemical Engineering Science* 56: 97-108.

Lyford P.A., Shallcross D. C., Grieser F., Pratt H.R.C. (1998). The Marangoni Effect and Enhanced Oil Recovery Part 2. Interfacial Tension and Drop Instability. *Canadian Journal of Chemical Engineering* 76:198-217.

Marangoni, C. G. M. (1865). Sull Expansiome dell Goccie di un Liquido Galleggianti sulla Superficie di Altro Liquido. Tipografia del Fratelli Fusi, Pavia.

Marangoni, C. G. M. (1871).Ueber die Ausbreitung der Tropfen einer Flussigkeit auf der Oberflache einer anderen, *Ann. Phys. Chem. (Poggendorff)*, 143 (7): 337 - 354

Maroudas N. G. and Sawistowski H. (1964). Simultaneous Transfer of Two Solutes Across Liquid-Liquid Interfaces. *Chemical Engineering Science* 19: 919-931.

Mendes-Tatsis M.A. (2000). Enhanced Mass Transfer in Liquid-Liquid Systems. *Proceedings Int. Symposium on Multiphase Flow and Transport Phenomena*, ICHMT- MFTP-2000, 5-10 November 2000, Antalya, Turkey, Keynote Lecture 12-01, Ed. David Moalem Maron.

Mendes-Tatsis, M. A., and Perez de Ortiz, E. S. (1992). Spontaneous Interfacial Convection in Liquid-Liquid Binary Systems Under Microgravity, *Proceedings of the Royal Society London*, A438, 389-396.

Mudge, L. K., and Heideger, W. J. (1970). The Effect of Surface Active Agents on Liquid-Liquid Mass Transfer Rates. *Journal of the American Institute of Chemical Engineers*, 16 (4): 602-608.

Nakache E., Dupeyrat M., Vignes-Adler M. (1983). Experimental and Theoretical Study of an Interfacial Instability at Some Oil-Water Interfaces Involving a Surface-Active Agent: I. Physicochemical Description and Outlines for a Theoretical Approach. *Journal of Colloid and Interface Science*. 94(1): 120-127.

Nakache E., and Raharimalala S. (1988). Interfacial Convection Driven by Surfactant Compounds at Liquid Interfaces: Characterisation by a Solutal Marangoni Number. In: Velarde M G, editor. *Physicochemical Hydrodynamics: Interfacial Phenomena*. Plenum Press, New York and London,

Nakache E., Raharimalala S., Vignes-Adler M. (1991). Marangoni effect in Liquid-Liquid Extraction with Surface Active Agent. In: G F Hewitt, F. Mayinger and J R Riznic, editors. *Phase-Interface Phenomena in Multiphase Flow*, Hemisphere Publications Corporation, London, 573-582.

Orell A. and Westwater J.W. (1962). Spontaneous Interfacial Cellular Convection Accompanying Mass Transfer: Ethylene Glycol - Acetic Acid - Ethyl Acetate. *Journal of the American Institute of Chemical Engineers* 8(3): 350-356.

Perez de Ortiz, E. S. (1991). Marangoni Phenomena. In *Science and Practice of Liquid-Liquid Extraction*, Ed. Thornton, J.D., Clarendon Press.

Perez de Ortiz, E. S., and Sawistowski, H. (1973). Interfacial Stability of Binary Liquid-Liquid Systems - I. Stability Analysis, *Chemical Engineering Science* 28: 2051-2061.

Perez de Ortiz E. S. and Sawistowski H. (1973). Interfacial Stability of Binary Liquid-Liquid Systems-II. Stability Behaviour of Selected Systems. *Chemical Engineering Science* 28: 2063-2069.

Pursell, M. R. and Mendes M. A. (2003) Manuscript under preparation.

Pursell, M. R., Mendes-Tatsis M. A., and Stuckey D. C. (2000). The Effect of Surfactants During Solvent Extraction of Erythromycin-A from Buffer and Filtered Fermentation Broth, *Solvent Extraction for the 21st Century (Proceedings of ISEC'99)* Eds: M. Cox, M. Hidalgo and M. Valiente, Society of Chemical Industry, London, 155 –161.

Pursell, M. R., Mendes-Tatsis M. A., and Stuckey D. C. (2003a). Co-Extraction During Reactive Extraction of Phenylalanine using Aliquat 336 – Modelling Extraction Equilibrium, *Biotechnology and Bioengineering* 82 (5): 533-542.

Pursell, M. R., Mendes-Tatsis M. A., and Stuckey D. C. (2003b). Co-Extraction During Reactive Extraction of Phenylalanine using Aliquat 336: Interfacial Mass Transfer, *Biotechnology Progress* 19: 469-476.

Pursell, M. R., Mendes-Tatsis M. A., and Stuckey D. C. (2003c). The Effect of Fermentation Broth and Biosurfactants on Mass Transfer During Liquid-Liquid Extraction accepted for publication, *Biotechnology and Bioengineering* .

Sanfeld A. and Steinchen A. (1984). Chemical Instabilities In: Nicolis G, Baras F, editors. *NATO Adv. Science Institute, Series C, D*, 199.

Sawistowski, H. (1971). Interfacial Phenomena. In *Recent Advances in Liquid-Liquid Extraction*, Ed. Hanson, C., Pergamon Press, 293-365.

Sherwood T. K., Wei J. C. (1957). Interfacial Phenomena in Liquid Extraction. *Industrial and Engineering Chemistry* 49(6): 1030-1034.

Slavchev, S. and Mendes, M. A. (2003). Marangoni Instability in Binary Liquid-Liquid Systems. Submitted to Int. J. Heat & Mass Transfer.

Slavchev, S., Kalitzova-Kurteva, P. and Mendes, M. A. (2003). Manuscript under preparation.

Sörensen T.S. (1979). Instabilities induced by Mass Transfer, Low Surface Tension and Gravity at Isothermal and Deformable Fluid Interfaces In *Lecture Notes in Physics: Dynamics and Instability of Fluid Interfaces*: Springer-Verlag, 105: 1-74.

Sternling C. V., and Scriven L. E. (1959). Interfacial Turbulence Hydrodynamic Instability and the Marangoni Effect, *Journal of the American Institute of Chemical Engineers*, 5 (4): 514-523.

Thompson, J. (1855). On Certain Curious Motions Observable at the Surfaces of Wine and Other Alcoholic Liquors, *Philosophical Magazine*, 10: 330-335.

Van Voorst Vader F. (1960). Adsorption of Detergents at the Liquid-Liquid Interface. Part 1. *Transactions of the Faraday Society* 56: 1067-1077.

Characterisation of Adsorption Layers at Liquid Interfaces - Studies with drop and bubble methods

Reinhard Miller[1] and Valentin Fainerman[2]

[1] Max-Planck-Institut für Kolloid- und Grenzflächenforschung, Am Mühlenberg 1, D-14424 Potsdam/Golm, Germany

[2] Medical Physicochemical Centre, Donetsk Medical University, 16 Ilych Avenue, 83003 Donetsk, Ukraine

1   Introduction

The non-equilibrium properties of interfacial layers have a large impact on various technologies, comprising food processing, coating, oil recovery, and in particular the formation and stabilisation of foams and emulsions in widespread fields of application.

Theoretical models have reached a state that allows a quantitative description of the equilibrium state by thermodynamic models, the adsorption kinetics of surfactants at fluid interfaces, the transfer across interfaces and the response to transient or harmonic perturbations. As result adsorption mechanisms, exchange of matter mechanisms and the dilational rheology are obtained. For some selected surfactant systems, the characteristic parameters obtained on the various levels coincide very well so that a comprehensive understanding was reached.

This contribution comprises an overview for the three main fields of fundamental understanding of surfactants at liquid interfaces, and examples for experimental methodologies suitable for their study including results for selected surfactant systems:

- Thermodynamics of Surfactant Adsorption
- Kinetics of Surfactant Adsorption
- Dynamic Surface Tension of Solutions
- Drop and Bubble Shape Experiments
- Adsorption Behaviour of Mixed Systems

In general one can say that the thermodynamic description of an adsorption layer at a liquid interface provides the basis for the dynamic and mechanical understanding. As it is the final state of a process, it controls also the mechanism of its formation, the adsorption kinetics (sf. Fig. 1). The response to small or large deformations of a liquid interface is governed by the adsorption mechanism and hence the thermodynamic characteristics. After a compression, the surface concentration Γ reaches values larger than the respective equilibrium adsorption Γ_0 and a desorption process sets in. Both, adsorption and desorption induced by interfacial perturbations are processes governed by the thermodynamic and kinetic characteristics. Thus, the surface rheological behaviour seems to be most sensitive to the specificity of adsorbed surfactants.

The schematic summarises the interrelations between the three fields for a comprehensive macroscopic description of surfactant layers, however, neglects other characteristics such as molecular structure, electrical or magnetic properties

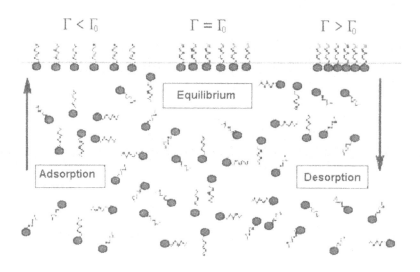

$\Gamma < \Gamma_0$ $\Gamma = \Gamma_0$ $\Gamma > \Gamma_0$

Equilibrium

Adsorption   Desorption

Figure 1. Schematic of processes and states of an interfacial surfactant layer and the adjacent solution bulk phase

2   Thermodynamics of Surfactant Adsorption

The thermodynamics of adsorption allows describing the relationships between measurable quantities in order to determine the adsorbed amount of surfactants at interfaces. It is also the basis of all models for adsorption dynamics and exchange of matter as response to interfacial perturbations. After a general introduction into the basic ideas of surface thermodynamics, classical as well as recently developed isotherms are presented and discussed. The models considering orientational changes of adsorbed molecules or formation of two-dimensional aggregates at the interface appear as generalisation of the fundamental models by Langmuir or Frumkin. A detailed discussion of experimental examples demonstrates the physics of the model parameters obtained from measured surface tension isotherms (Fainerman et al. 2001).

2.1 Fundamental Relationships

The chemical potentials of adsorbed components μ_i^s depend on the composition of the layer and its surface tension γ and is given by the relationship (Defay and Prigogine 1966, Rusanov 1967)

$$\mu_i^s = \mu_i^{0s}(T,P,\gamma) + RT \ln f_i^s x_i^s \tag{1}$$

where $\mu_i^{0s}(T, P, \gamma)$ is the standard chemical potential of component i and depends on temperature T, pressure P and surface tension γ, and the f_i are the activity coefficients. Here the superscript 's' refers to the surface (interface). The standard chemical potential can be presented as a function of pressure and temperature only, if one introduces an explicit dependence of $\mu_i^{0s}(T, P, \gamma)$ on surface tension into Eq. (1). To do this, the well-known expression for the variation of the free enthalpy of the Gibbs dividing surface at constant pressure and temperature (Defay and Prigogine 1966, Rusanov 1967) can be used

$$dG = -Ad\gamma + \sum \mu_i^s dm_i^s ,$$ (2)

where A is the surface area. As dG is a total differential, each component at the surface obeys the following Maxwell relationship

$$\left(\frac{\partial \mu_j^s}{\partial \gamma} \right)_{m_j^s} = - \left(\frac{\partial A}{\partial m_j^s} \right)_\gamma$$ (3)

at constant p, T and numbers of molecules except j. The derivative on the right hand side of Eq. (2.4) is by definition the partial molar area of the jth component (index 0 refers to the pure solvent water)

$$\left(\frac{\partial A}{\partial m_j^s} \right)_\gamma = \omega_j$$ (4)

at constant p, T and numbers of molecules other than j. Using Eqs. (3) and (4) one can transform Eq. (1) into the form

$$\mu_i^s = \mu_i^{0s}(T, P) - \int_0^\gamma \omega_i d\gamma + RT \ln f_i^s x_i^s .$$ (5)

In contrast to Eq. (1), here the standard chemical potential $\mu_i^{0s}(T, P) = \mu_i^{0s}$ is already independent of surface tension. Assuming that ω_i is also independent of γ, and integrating Eq. (5), one obtains the expression

$$\mu_i^s = \mu_i^{0s} + RT \ln f_i^s x_i^s - \gamma \omega_i, \tag{6}$$

which is called the Butler equation (1932), which is often used to derive surface equations of state and adsorption isotherms. The equation is to be applied to a Gibbs dividing surface chosen by a convention which results in positive values for all adsorptions, including that of the solvent. Such a convention can be formulated by choosing numerical values for the partial molar surface areas of all components, as first recognised by Joos (1967).

Equations of state for surface layers, adsorption isotherms and surface tension isotherms can be derived by equating the expressions for the chemical potentials at the surface, Eq. (6), to those in the solution bulk. As shown for example in (Fainerman et al. 2001) for ideally dilute solutions this finally yields

$$\ln \frac{f_0^s x_0^s}{f_0^\alpha x_0^\alpha} = -\frac{(\gamma_0 - \gamma)\omega_0}{RT}. \tag{7}$$

and

$$\ln \frac{f_i^s x_i^s / f_{(0)i}^s}{K_i f_i^\alpha x_i^\alpha / f_{(0)i}^\alpha} = -\frac{(\gamma_0 - \gamma)\omega_i}{RT}. \tag{8}$$

For a solution that contains only one surfactant, and assuming $\omega_0 = \omega_1$ the Eqs. (7) and (8) transform into the well-known equations of von Szyszkowski (1908) and Langmuir (1907)

$$\Pi = \gamma_0 - \gamma = \frac{RT}{\omega_1} \ln(1 + b_1 c_1), \tag{9}$$

$$\Gamma_1 = \frac{1}{\omega_1} \frac{b_1 c_1}{1 + b_1 c_1}, \tag{10}$$

respectively, were the constant b_1 is the surface-bulk distribution coefficient related to the concentration c rather than to the mole fraction x.

It is seen from the von Szyszkowski-Langmuir surface tension isotherm, that at a given temperature the shape of the surface tension isotherm is determined only by the parameter ω, while parameter b enters this equation as a dimensionless variable bc. The dependence of the surface pressure isotherm on the molar area ω is illustrated by Fig. 2. For lower ω steeper curves $\Pi(c)$ are obtained.

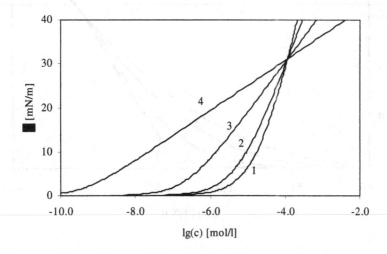

Figure 2. Surface pressure as a function of bulk concentration for the Szyszkowski-Langmuir model Eqs. (9), (10) with $\omega_1 = 1.5 \cdot 10^5$ m²/mol (1), $2.5 \cdot 10^5$ m²/mol (2), $5.0 \cdot 10^5$ m²/mol (3) and 10^6 m²/mol (4).

The Frumkin equation of state and adsorption isotherm involve an additional interaction parameter a (for a = 0 we obtain the Langmuir model),

$$\Pi = -\frac{RT}{\omega_0}\left[\ln(1-\theta) + a\theta^2\right],$$ (11)

$$bc = \frac{\theta}{(1-\theta)}\exp(-2a\theta).$$ (12)

The Frumkin model can better fit experimental data due to this additional parameter. The effect of the parameter a at fixed ω is illustrated by Fig. 3.

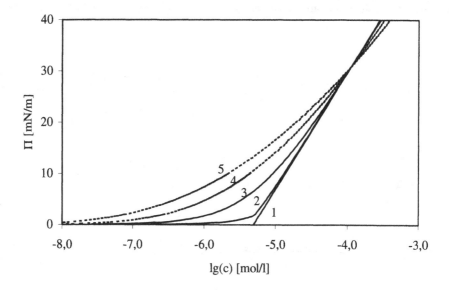

Figure 3   Frumkin isotherms calculated from Eqs. (11) and (12), with ω = 2.5·105 m²/mol and: a = 4 (1), a = 2 (2), a = 0 (3), a = -2 (4), a = -4 (5).

For negative values of a (which can result from intermolecular repulsion in the surface layer, which is the case for ionic surfactants) the shape of the curve is steeper as compared with the Langmuir isotherm, while for positive a-values (intermolecular attraction in the monolayer) the curve becomes steeper at low concentrations, and is almost a straight line at higher concentrations. All the curves shown in Figs. 2 and 3 are normalised with respect to the constant b in such a way that the concentration 10^{-4} mol/l corresponds to a surface pressure of 30 mN/m.

2.2 New Adsorption Layer Models

The fact that many surfactant systems cannot be adequately described by the Frumkin mode was the reason that other models have been derived. A comprehensive overview of such models was given recently elsewhere (Fainerman et al. 1998). We want to discuss two of the most recent models considering changes in orientation of adsorbed molecules and formation of two-dimensional aggregates (Fainerman et al. 2002). These new models are suitable to describe quite a number of surfactant adsorption layers much better than classical models do.

2.2.1 Surface layers formed by surfactants able to change orientation

In this model it is assumed that a reorientation of adsorbed molecules leads to a variation of the partial molar area ω_i. Taking this possibility into account the following set of equations is obtained (Fainerman et al. 1996, 1997, 2003)

$$-\frac{\Pi\omega_0}{RT} = \ln(1-\Gamma_\Sigma\omega) + \Gamma_\Sigma\omega - \Gamma_\Sigma\omega_0 + a(\Gamma_\Sigma\omega)^2 \qquad (13)$$

$$b_i c = \frac{\Gamma_i\omega_0}{(\omega_i/\omega_1)^\alpha (1-\Gamma_\Sigma\omega)^{\omega_i/\omega_0}} \exp\left(-\frac{\omega_i}{\omega_0}(2a\Gamma_\Sigma\omega)\right) \qquad (14)$$

Here $\Gamma_\Sigma = \sum_{i\geq 1}\Gamma_i$ is the total adsorption in all n states, α is a constant, and ω is the average molar area of all states calculated from the relationship $\omega = (\sum_{i_{min}}^{i_{max}} \omega_i\Gamma_i)/\Gamma_\Sigma$, with $i_{min} = \omega_{min}/\omega_0, i_{max} = \omega_{max}/\omega_0$, and n = $i_{max}-i_{min}$. The distribution of adsorptions over the states is then given by the expressions

$$\Gamma_i = \Gamma_\Sigma \frac{\left(\frac{\omega_i}{\omega_1}\right)^\alpha (1-\Gamma_\Sigma\omega)^{\frac{\omega_i-\omega_1}{\omega_0}} \exp\left[\left(\frac{\omega_i-\omega_1}{\omega_0}\right)(2a\Gamma_\Sigma\omega)\right]}{\sum\limits_{i_{min}}^{i_{max}} \left(\frac{\omega_i}{\omega_1}\right)^\alpha (1-\Gamma_\Sigma\omega)^{\frac{\omega_i-\omega_1}{\omega_0}} \exp\left[\left(\frac{\omega_i-\omega_1}{\omega_0}\right)(2a\Gamma_\Sigma\omega)\right]} \qquad (15)$$

Assuming that a = 0, one can rewrite Eqs. (13) and (14) as:

$$-\frac{\Pi\omega_0}{RT} = \ln(1-\Gamma_\Sigma\omega) + \Gamma_\Sigma\omega - \Gamma_\Sigma\omega_0 \qquad (16)$$

$$bc = \frac{\Gamma_1\omega_0}{(1-\Gamma_\Sigma\omega)^{\omega_1/\omega_0}} \qquad (17)$$

where we assumed i = 1 in Eq. (17) for definiteness.

Another method for the solution of Eqs.(7) and (8) for reorienting surfactant molecules can be used. It is possible to assume that the molar areas of the solvent and surfactant are equal. I.e., for two states with the minimal and maximal molar area, $\omega_0 = \omega = (\omega_1\Gamma_1 + \omega_2\Gamma_2)/\Gamma_\Sigma$. In this case the contribution of the non-ideality of entropy vanishes. Then, the adsorption isotherms for molecules existing in two states (1 and 2) in an ideal (with respect to enthalpy) surface layer become:

$$-\frac{\Pi\omega}{RT} = \ln(1 - \Gamma_\Sigma\omega) \tag{18}$$

$$bc = \frac{\Gamma_1\omega}{(1 - \Gamma_\Sigma\omega)^{\omega_1/\omega}} \tag{19}$$

Here the ratio of adsorptions in the state with minimum (ω_1) and maximum (ω_2) molar areas is expressed by a relation which follows from Eq.(15):

$$\frac{\Gamma_2}{\Gamma_1} = \exp\left(\frac{\omega_2 - \omega_1}{\omega}\right)\left(\frac{\omega_2}{\omega_1}\right)^\alpha \exp\left[-\frac{\Pi(\omega_2 - \omega_1)}{RT}\right] \tag{20}$$

It was noted above that the choice of $\omega_0 \neq$ const, is incompatible with Eq. (6); however, for practical purposes this may be justified when the difference $\omega_2 - \omega_1$ is small as compared to ω.

Some model calculations using Eqs. (18)-(20) are summarised in Fig. 4. The coefficient α was set zero and b was again adjusted such that the concentration 10^{-4} mol/l corresponds to a surface pressure of 30 mN/m. For $\omega_1 = \omega_2$ we get the curve for the Langmuir model (curve 2 – dotted line).

As an example, the surface tension isotherms for alkyl dimethyl phosphine oxides (C_nDMPO) for C_8 to C_{16} at 25 °C are shown in Fig. 5. The Frumkin and reorientation models agree both well with the experimental data. Small differences between the calculated isotherms exist only for $n \geq 13$, while for lower n the curves for the two models perfectly coincide. Thus, neither of the two models can be preferred if one takes into account only the agreement between the experimental and theoretical data. However, negative values of the Frumkin constant a are obtained for lower homologues, and an unexpected dependence of a(n) are obtained, which together indicate that the coincidence between the Frumkin model and the experimental data is only formal.

The dependence of molar area for the two states of C_nDMPO molecules on the number of carbon atoms in the hydrocarbon chain is shown in Fig. 6. While the minimum area per C_nDMPO

mole (or per molecule) ω_2 is almost independent of n, the area per mole (or molecule) in the unfolded state ω_1, increases with the chain length for $n \geq 12$.

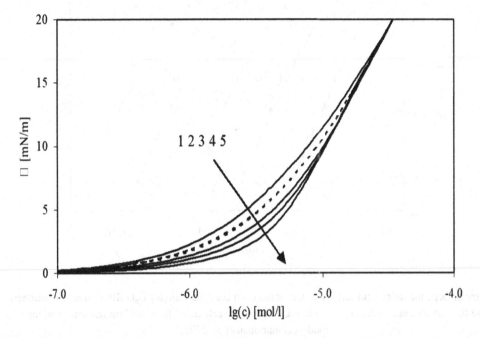

Figure 4. Dependence of surface pressure on the bulk concentration for the following parameter values: $\omega_2 = 2.5 \cdot 10^5$ m²/mol; $\omega_1 = 5 \cdot 10^5$ m²/mol (1), $2.5 \cdot 10^5$ m²/mol (2) (corresponds to the Langmuir model), $1.5 \cdot 10^6$ m²/mol (3), $2.0 \cdot 10^6$ m²/mol (4) and $4.0 \cdot 10^6$ m²/mol (5).

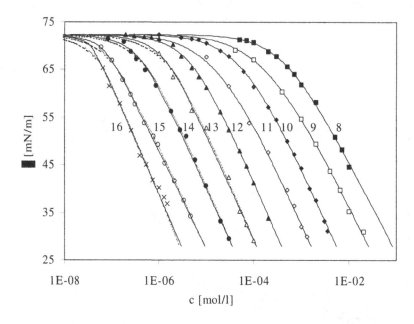

Figure 5. Dependence of equilibrium surface tension on concentration for C_nDMPO solutions, numbers denote the carbon atoms number, the theoretical curves are calculated from the Frumkin and reorientation models (Fainerman et al. 2001)

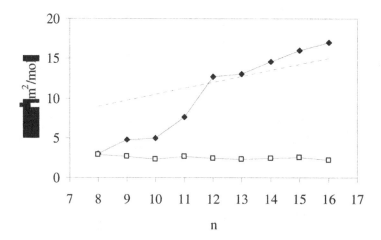

Figure 6. Dependence of molar area in the states 1 ($\blacklozenge$) and 2 ($\square$) on the number of carbon atoms n for the C_nDMPO, dashed line - estimated from molecular geometry.

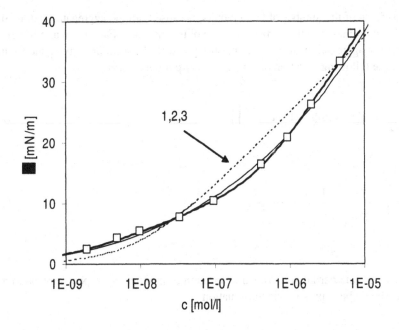

Figure 7. Dependence of equilibrium surface pressure on the concentration of $C_{14}EO_8$; ($\square$) - data from Ueno et al. (1981) at 25°C. Theoretical curves were calculated in (Fainerman et al. 2003) from the Langmuir models (curve 1), and reorientation models (15)-(17) and (18)-(20), curves 2 and 3 respectively.

It is interesting to note that the area per mole C_nDMPO calculated from the atomic radii and bond lengths for the state with maximum area agrees well with the data obtained from tensiometry for $n > 11$. For lower C_nDMPO homologues the molecule in a state flat at the interface is not advantageous.

In Figure 7 the experimental data by Ueno et al. (1981) for $C_{14}EO_8$ solutions are compared to the surface pressures calculated from the Langmuir model (11)-(12), and both reorientation models: (15)-(17) and (18)-(20). The best agreement with the experiments is given by the reorientation models, while the deviations for Langmuir models is quite large, not surprisingly perhaps in view of the cruder nature of the latter models.

2.2.2 Surfactant adsorption layers with two-dimensional aggregation

The aggregation of surfactant molecules in the solution bulk, e.g., micelle formation, affects significantly the shape of surface tension isotherms. For a non-ionic surfactant solution with $c > CMC$, the Gibbs adsorption equation has the form $d\gamma/d\ln c \cong -RT\Gamma/n$, where n is the micellar aggregation number, usually between 50 and 100 (Rusanov and Fainerman 1989). The mechanism, however, through which surfactants aggregate in the surface layer is different. While aggregation in the bulk leads to the formation of surface inactive micelles and hence to a decrease of the concentration of surface active monomers in the bulk, the surface aggregation of surfactants leads to a decrease of the number of kinetic units, because both the aggregates and the monomers have a similar contribution to the surface pressure of the interface (Fainerman and Miller 1996).

The model for aggregation in the adsorption layer with the arbitrary aggregation number n is described by the following equation of state and adsorption isotherm

$$-\frac{\Pi\omega}{RT} = \ln\left\{1 - \Gamma_1\omega\left[1 + (\Gamma_1/\Gamma_c)^{n-1}\right]\right\} \tag{21}$$

$$bc = \frac{\Gamma_1\omega}{\left\{1 - \Gamma_1\omega\left[1 + (\Gamma_1/\Gamma_c)^{n-1}\right]\right\}^{\omega_1/\omega}}. \tag{22}$$

Here Γ_1 and Γ_c are the partial and critical adsorption of monomers, respectively, and the average molar area ω should be expressed via the equation

$$\frac{\omega}{\omega_1} = \frac{1 + n(\Gamma_1/\Gamma_c)^{n-1}}{1 + (\Gamma_1/\Gamma_c)^{n-1}}. \tag{23}$$

To demonstrate the effect of the model parameters on the shape of the surface pressure isotherms, the process of dimer formation in the adsorption layer at various Γ_c is shown in Fig. 8. The smaller Γ_c is, the more pronounced is the difference between the aggregation and Langmuir isotherm (n = 1). For $\Gamma_c < 10^{-10}$ mol/m² the shape of the surface pressure isotherm becomes independent of Γ_c because the adsorption layer contains only aggregates.

Also the aggregation number has a direct influence on the isotherm shape, however, for n > 20 it becomes also independent of n. If the critical adsorption Γ_c is sufficiently large, the curves exhibit a characteristic kink which indicates the formation of clusters in the adsorption layer.

As example, the results for some fatty acids in acidic solution are illustrated in Fig. 9. One can see that for C_7–C_9 the corresponding theoretical curves are indistinguishable, i.e. both theoretical models provide good agreement. However, for decanoic and lauric acids the aggregation model describes the data better. The slope of the dependence b(n) for fatty acids and normal alcohols is essentially the same (Fainerman et al. 2002), however, the values for the fatty acids are 1.5 times lower (cf. Fig. 10).

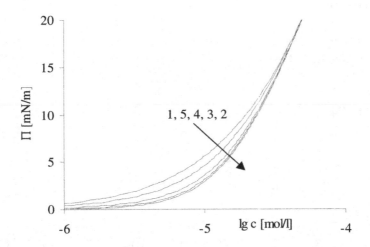

Figure 8. Surface pressure isotherm for a surfactant solution, calculated from the Langmuir equation for monomers (1), and for the formation of surface dimers: $\Gamma_c = 10^{-10}$ mol/m² (2); $\Gamma_c = 10^{-8}$ mol/m² (3); $\Gamma_c = 10^{-7}$ mol/m² (4) and $\Gamma_c = 10^{-6}$ mol/m² (5); ω_1 is varied in the range $(1-1.5) \cdot 10^5$ m²/mol to obtain the same slope of all curves at $\Pi > 20$ mN/m; the constant b was chosen such that $\Pi = 30$ mN/m was reached at a concentration of 10^{-4} mol/l.

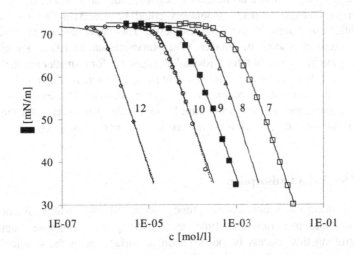

Figure 9. Dependence of equilibrium surface tension on concentration for heptanoic and nonanoic acids (Malysa et al. 1991, 20°C), octanoic acid (Aratono et al. 1984, 25°C), decanoic acid (Lunkenheimer and Hirte 1992, 25°C), and dodecanoic acid (Lucassen-Reynders 1972, 20°C), numbers denote the number of carbon atoms, the theoretical curves are calculated from the Frumkin and aggregation models, respectively.

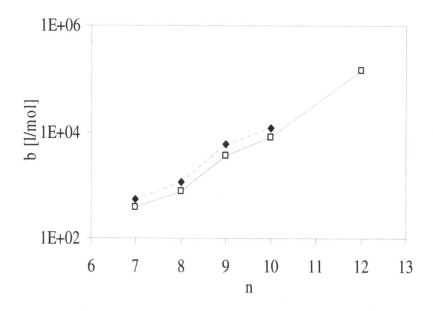

Figure 10. Dependence of adsorption equilibrium constant b on n for the fatty acids series ($\square$), data for alcohols ($\blacklozenge$)

For the dodecanoic acid, in contrast to 1-dodecanol, the aggregation model yields a quite high aggregation number of n = 4, while no indications exist for the formation of larger clusters. Also, no data for spread dodecanoic acid monolayers could be found in literature which exhibit characteristic inflection points in the Π-A isotherm. The dependencies at 15-20 °C usually correspond to the type of gaseous monolayers (Ter Minassian-Saraga 1956). Therefore, a certain similarity exists in the behaviour of fatty acids and alcohols, i.e. for a smaller number of methylene groups the system is better described by the Frumkin model, with increasing number of methylene groups the behaviour of the system becomes more 'aggregation-like'. However, as the solubility of the acids is higher and therefore the b values are lower, condensation sets in for the acids only at C_{13} (Boyd 1958), but not at C_{12}, which is characteristic to the normal alcohols.

3 Kinetics of Surfactant Adsorption

The adsorption of surfactants at interfaces is a time process. After the creation of a new surface the adsorption is zero and increases with time until reaching the equilibrium state. The main mechanism controlling this process is the diffusion of surfactants in the solution bulk. In this lesson the basic models will be discussed and the main physical parameters analysed. In particular, the type of adsorption isotherm plays an important role. On the basis of dynamic surface tensions the application of the theoretical models will be demonstrated in the subsequent paragraph. Besides complete solutions of the diffusion model, also approximate solutions exist. These models have different ranges of application and are easy to handle. However, with the use of a suitable software, also the complete theories can be applied to experimental data without the need of complicated mathematical procedures.

The present state of research allows describing the adsorption kinetics for most surfactants at liquid interfaces quantitatively. The first model derived by Ward & Tordai (1946) was based on the assumption that the time dependence of interfacial tension, which is directly related to the interfacial concentration Γ of adsorbed molecules via an equation of state, is mainly caused by the diffusional transport of surfactant molecules to the interface. A schematic picture of this model is shown in Fig. 1. Transport in the solution bulk is controlled by surfactant diffusion. The transfer of molecules from the so-called subsurface, the liquid layer adjacent to the interface, to the interface itself is assumed to happen without transport.

The adsorption of surface active molecules at an interface is a dynamic process. In equilibrium the adsorption flux to and desorption flux from the interface are in balance. If the actual surface concentration is smaller than the equilibrium one, $\Gamma < \Gamma_0$, the adsorption flux predominates, if $\Gamma > \Gamma_0$, the desorption flux prevails. If processes happen in the adsorption layer, such as changes in the orientation or conformation of adsorbed molecules, or the formation of aggregates due to strong intermolecular interaction, additional fluxes within the adsorption layer have to be considered.

Milner (1907) was the first to discuss diffusion as the process responsible for the time dependence of surface tension of soap solutions. Later, several models took into account transfer mechanisms in form of rate equations (Doss 1939, Hansen and Wallace 1959). More complicated models accounted for diffusion and transfer mechanisms simultaneously (Baret 1969, Miller and Kretzschmar 1980, Ravera et al. 1994).

3.1 The physical model

Most quantitative descriptions of adsorption kinetics processes are so far based on the model derived in 1946 by Ward and Tordai. The model assumes that the step of transfer from the subsurface to the interface is fast compared to the transport between the bulk and the subsurface by diffusion. If any flow in the bulk phase is neglected diffusion is described by Fick's second diffusion law

$$\frac{\partial c}{\partial t} = D\frac{\partial^2 c}{\partial x^2} \quad \text{at } x > 0, t > 0. \tag{24}$$

A suitable boundary condition is Fick's first law defined at the surface located at x=0,

$$\frac{\partial \Gamma}{\partial t} = j = D\frac{\partial c}{\partial x} \quad \text{at } x = 0; t > 0. \tag{25}$$

To complete the transport problem one additional boundary condition

$$\lim_{x \to \infty} c(x,t) = c_o \text{ at } t > 0, \tag{26}$$

and an initial condition are needed, for example a homogenous concentration distribution and a freshly formed interface with no surfactant adsorption, respectively,

$$c(x,t) = c_o \text{ at } t = 0, \tag{27}$$

$$\Gamma(0) = 0. \tag{28}$$

The solution can be found by applying the Laplace operator method (Hansen 1961), leading to two equivalent relationships,

$$\Gamma(t) = 2\sqrt{\frac{D}{\pi}} \left(c_o \sqrt{t} - \int_0^{\sqrt{t}} c(0, t - \tau) d\sqrt{\tau} \right) \tag{29}$$

and

$$c(0,t) = c_o - \frac{2}{\sqrt{D\pi}} \int_0^{\sqrt{t}} \frac{d\Gamma(t - \tau)}{dt} d\sqrt{\tau}. \tag{30}$$

Here D is the diffusion coefficient, c_o is the surfactant bulk concentration, $c(x,t)$ is the surfactant concentration as a function of time t and distance to the interface x. The integral equation describes the change of Γ with time, and can be applied to dynamic surface tension data $\gamma(t)$. Its use, however, requires an additional equation, a surface tension isotherm.

As shown above, various such equations exist, such as the classical ones named Langmuir or Frumkin isotherm. However, it was also shown that peculiarities of surfactants in adsorption layers can be described quantitatively only if special models are used. Their impact on adsorption kinetics was reviewed recently (Fainerman et al. 1998) and found to be significant. While the reorientation of adsorbed molecules mimics an acceleration of the adsorption process, surface aggregation on the contrary apparently decelerates it.

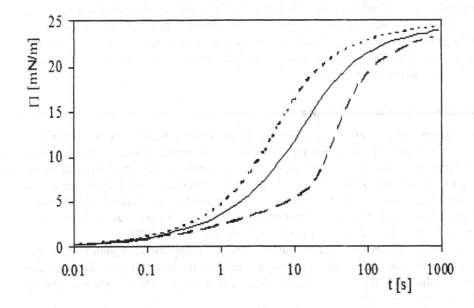

Figure 11. Dynamic surface pressure calculated for a Frumkin isotherm; $c_0 = 5 \cdot 10^{-8}$ mol/cm^3, $D = 6 \cdot 10^{-10}$ m^2/s, $\omega = 1.5 \cdot 10^5$ m^2/mol; a = 0 (refers to a Langmuir isotherm); a = 1.5 – dotted line; a = -1.5 – dashed line; the parameter b was chosen such that the equilibrium surface tension is 25 mN/m

Based on a linear isotherm $\Gamma(t) \sim c(0,t)$, a simple equation was derived by Sutherland (1952)

$$\Gamma(t) = \Gamma_o \left(1 - \exp(Dt/K^2)\, \mathrm{erfc}(\sqrt{Dt}/K)\right).$$ (31)

Its range of application however is very limited (Miller 1990). Via the Gibbs fundamental equation

$$\Gamma(t) = -\frac{1}{RT}\frac{d\gamma(t)}{d\ln c(0,t)},$$ (32)

a relationship for the dynamic interfacial tension results,

$$\gamma(t) = \gamma_o - RT\Gamma_o \left(1 - \exp(Dt/K^2)\mathrm{erfc}(\sqrt{Dt}/K)\right).$$ (33)

To demonstrate the influence of the adsorption isotherm on the adsorption kinetics of a surfactant, the change in surface tension with time $\gamma(t)$ for a Frumkin isotherm is shown in Fig. 11 for three different interaction parameters a of Eq. (11). As one can see, the shape of the adsorption isotherm has a significant influence on the course of the adsorption kinetics, here given in terms of dynamic surface pressure as a function of time.

3.2 Consideration of interfacial reorientation

In the thermodynamic model derived by Fainerman et al. (1996), which has been described in detail above, it was supposed that, depending on the surface coverage, surfactant molecules can adsorb in two different states. These states are characterised respectively by two different molar surface areas ω_1 and ω_2, and by the parameters b_1 and b_2, which are related to the respective surface activities. Surfactants like n-alkyl dimethyl phosphine oxides (Aksenenko et al. 1998) and poly-oxethylated alcohols adsorbed at the water/air or water/alkane interfaces can be described by this model perfectly (Ferrari et al. 1998, Liggieri et al. 1999, Miller et al. 2000).

In order to describe the evolution of the surface pressure during the adsorption process, the diffusion process in the bulk, and the change in the orientation of the adsorbed molecules will have to be considered, which was discussed in detail by Ravera et al. (2000). The characteristic time of the orientation process is given by

$$\tau_{or} = \frac{1}{k_{12}\left[1 + \dfrac{\Gamma_{1,0}}{\Gamma_{2,0}}\right]} \tag{34}$$

which depends both on the orientation rate constant k_{12} and the partitioning between the two adsorption states Γ_1/Γ_2 at equilibrium. The comparison of this characteristic time with the diffusion relaxation time τ_D

$$\tau_D = \frac{1}{D}\left(\frac{\Gamma_0}{c_0}\right)^2 \tag{35}$$

can be used to verify which process governs the adsorption dynamics.

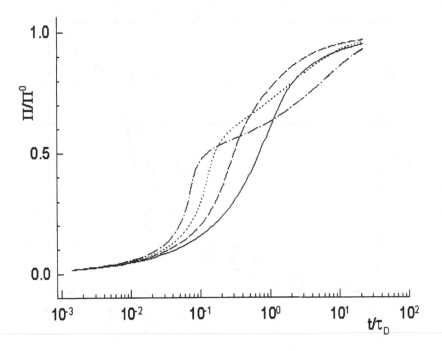

Figure 12. Influence of the ratio ω_2/ω_1 on the dimensionless surface pressure Π/Π_0 versus dimensionless time t/τ_D, $K_{12}=0.2$ (corresponding to $\tau_D/\tau_{or} \approx 1$); $c_0=6 \cdot 10^{-8}$ mol/cm³, $b_2=1.4 \cdot 10^8$ cm³/mol, $\alpha=2.2$, $\omega_1=6.7 \cdot 10^9$ cm²/mol; $\omega_2/\omega_1= 1$ (—), 2 (---), 3 (·····), 4 (-·-), according to Ravera et al. (2000)

When the molecular orientation processes are at equilibrium with respect to the diffusional transport, $\tau_D \gg \tau_{or}$, the adsorption changes essentially due to the diffusive flux to the surface. Here it is assumed that the process of exchange of molecules between the interface and the sublayer (kinetic transfer) is at equilibrium in comparison to diffusion and orientation, i.e. $\tau_k \ll \tau_{or} \approx \tau_D$. Model calculations performed by Ravera et al. (2000) show the effect of the reorientation kinetics mechanism, which under certain conditions can be significant. As an example, results of calculations are given in Fig. 12 showing the influence of the ratio between the molar surface areas corresponding to the two adsorption states. The behaviour of the surface pressure strongly changes only when the ratio of ω_2/ω_1 exceeds a value of 3. This ratio is typical for surfactants of the C_nEO_m type. Note that $\omega_2/\omega_1 = 1$ corresponds to the Langmuir model. As one can see, in the beginning the reorientation process enhances the change in surface tension while at larger adsorption time the trend can change.

3.3 Consideration of interfacial aggregation

Above adsorption isotherms where discussed for surfactants able to form small aggregates (dimers or trimers) in the adsorption layer. There are quite a number of surfactants, which can be described by this model perfectly, for example the homologous series of fatty acids or alcohols or the alkyl sulphates (Fainerman et al. (2000)).

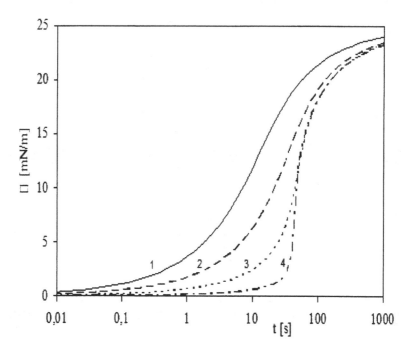

Figure 13. Dynamic surface pressure for an aggregation isotherm with $c_0 = 5 \cdot 10^{-8}$ mol/cm^3,
$D = 6 \cdot 10^{-6}$ cm^2/s, $\omega_1 = 1.5 \cdot 10^5$ m^2/mol; curve (1) corresponds to the Langmuir isotherm; aggregation
model: $\Gamma_c = 10^{-13}$ mol/cm^2, n=2 (2); n=5 (3); n=20 (4), the parameter b was chosen such that the
equilibrium surface tension is 25 mN/m

The modelling of adsorption kinetics of surfactants, which show interfacial aggregation has
been discussed recently elsewhere (Fainerman et al. 2000, 2000a Aksenenko 1998). To derive an
adsorption kinetics model the Ward and Tordai equation (29) is again the main relationship
between the dynamic adsorption $\Gamma(t)$ and the subsurface concentration c(0,t), and as adsorption
isotherm the equations (21) - (23). The set of equation is too complex to find an analytical
solution, but for the short time range and for low adsorption layer coverage, the following
approximation is valid (Aksenenko et al. 1998)

$$\Pi = \frac{2c_0RT}{n}\left(\frac{Dt}{\pi}\right)^{1/2} . \tag{36}$$

The aggregation number n is an additional parameter in the denominator, leading to a slower
adsorption caused by the formation of aggregates within the adsorption layer. For the whole time
interval, the set of equations together with the diffusion equation has to be solved numerically.

This model assumes that the aggregation process itself does not require additional time, i.e. there is always equilibrium between monomers and aggregates in the adsorption layer. To solve this set of equations numerically, first-order finite difference schemes can be applied as described in detail by Aksenenko et al. (1998a). The entire numerical procedure allows step-wise calculation of the time dependencies $\Gamma(t)$ or $\Pi(t)$.

As illustration the dependencies of dynamic surface pressure for a surfactant solution of $c_0 = 5 \cdot 10^{-8}$ mol/cm^3 and $D = 6 \cdot 10^{-6}$ cm^2/s assuming formation of small aggregates in the surface layer are shown in Fig. 13. The effect of the aggregation number on the dynamic surface pressure of a surfactant solution is very strong and leads to a sharp deceleration of the surface tension decrease for larger aggregation number.

3.4 Adsorption processes at interfaces of time variable area

Depending on the various types of experiments area deformations can happen or be produced as part of the method. In a very detailed analysis Joos (1999) has demonstrated that the adsorption to or desorption from a liquid interface with changing interfacial area can be described in a very general way. The respective diffusion controlled model reads

$$\Gamma f(t) = \Gamma_e + 2 \left(\frac{D\tau}{\pi}\right)^{1/2} c_0 - 2 \left(\frac{D}{\pi}\right)^{1/2} \int_0^{\sqrt{\tau}} c_s(c - \lambda) d\sqrt{\lambda} \qquad (37)$$

This equation has the same structure as the famous Ward and Tordai equation (29) discussed above. The function f(t) introduced by Joos

$$f(t) = \frac{A(t)}{A_0} \qquad (38)$$

represents the relative change of the interfacial area A(t) with time, A_0 is the initial area. Based on this model the effective interfacial age can be determined from

$$t_{eff} = \frac{\tau}{f^2(t)} \qquad (39)$$

with

$$\tau = \int_0^t f^2(t)dt = \int_0^t \left(\frac{A}{A_0}\right)^2 dt. \tag{40}$$

For data obtained by the maximum bubble pressure method, such a correction is needed as the bubble grows during the experiment.

3.5 Consideration of micelles in the bulk

When an adsorption layer is formed from a micellar solution, then monomers adsorb at the surface. This decreases the monomer concentration locally, the monomers and micelles are out of equilibrium and micelles will disintegrate. Hence, locally the concentration of micelles decreases and micelles will diffuse too. Thus, the presence of micelles in the solution bulk can be seen as extra source of matter, i.e. the micellar kinetics represents an additional relaxation mechanism to interfacial perturbations.

The coupling of the diffusion of monomers and micelles is given by the micellar kinetics, which consists of different physical processes: a fast process in the range of microseconds (exchange of monomers between the micellar and the aqueous solution phase), and a second in the range of milliseconds (total disintegration of micelles into monomers). The entire variety of micellar kinetics was discussed by Aniansson et al. (1976), and a quantitative model, considering the diffusion of monomers and micelles, and the micellar kinetics mechanisms, was reviewed in the paper by Dushkin (1998) or in the book by Joos (1999).

3.6 Effect of surfactants charge

The existence of an electric double layer can remarkably influence the dynamic interfacial properties of ionic surfactant solutions (van den Bogaert and Joos 1979, Fainerman et al. 1994). The equilibrium state of such interfacial layers has been described in much detail in Paragraph 2.5. The dynamic problems, however, are rather complex and difficulties arise in solving the respective set of non-linear equations.

First models derived by Dukhin et al. (1985), and Borwankar and Wasan (1988) used a quasi-equilibrium model by assuming that the characteristic diffusion time is much greater than the relaxation time of the electrical double layer, and thus, the complicated electro-diffusion problem is reduced to a simple transport problem. Datwani and Stebe (1999) analysed this model and performed extensive numerical calculations.

3.7 Adsorption and transfer across the interface

The description of adsorption processes at liquid-liquid interfaces requires some specific consideration. Almost each surfactant is soluble in both the aqueous and the oil phases which can practically never be neglected. This implies that a theoretical modelling of the adsorption kinetics has to consider the transfer of surfactant across the interface during the adsorption process. The initial partition state of surfactants have also to be carefully defined prior to the experiments, hence the ratio of volumes of the bulk phases becomes an important parameter for the adsorption dynamics (Ferrari et al. 1997, 1997a). For these reasons the knowledge of the distribution coefficient K for the adsorbing surfactant is a key parameter, which is defined as the ratio between the equilibrium surfactant concentrations in the two phases. For dilute and ideal solutions in the two liquids α and β, the ratio between the equilibrium concentrations c_α and c_β can be written as (Lyklema 1993)

$$K = \frac{c_\alpha}{c_\beta} = \frac{v_\alpha}{v_\beta} \exp\left(-\frac{\mu^{0\alpha} - \mu^{0\beta}}{RT}\right) \tag{41}$$

where v_α and v_β are the molar volumes. Although K describes properties of the liquid bulk phases, it is a key parameter in the adsorption dynamics of a surfactant at a liquid-liquid interface (Ferrari et al. 1997, Rubin and Radke 1980). There are only few attempts known from literature where K is determined. A straightforward procedure is the measurement of the bulk concentrations after equilibration of the two immiscible phases. This, however, only works for particular surfactants, and common analytic techniques usually fail at low surfactant concentration. An indirect method consists in the measurement of the surface tension of the aqueous phase. Via the γ-c isotherm as a master curve, the concentration can be evaluated then and hence the distribution coefficient (Ravera et al. 1997).

3.8 Approximate solutions

The use of the Ward and Tordai equation (29) requires large numerical efforts. Therefore, simple asymptotic and approximate solutions are very valuable and the approximations derived in 1952 by Sutherland in the form of Eqs. (31) or (33) have been very often used. In the literature a number of further approximation for small surface coverage and small and large adsorption times were discussed. Hansen (1960) for example expressed the change in surface concentration as a function of the dimensionless time $\Theta = Dt/(\Gamma_o/c_o)^2$

$$\frac{\Gamma(t)}{\Gamma_o} \approx 2\sqrt{\frac{\Theta}{\pi}}\left(1 - \frac{\sqrt{\pi\Theta}}{2} \pm ...\right) \text{ for } \sqrt{\Theta} \leq 0.2 \tag{42}$$

$$\frac{\Gamma(t)}{\Gamma_o} \approx 1 - \frac{1}{\sqrt{\pi\Theta}}\left(1 - \frac{1}{2\Theta} \pm ...\right) \text{ for } \sqrt{\Theta} \geq 5.0 . \tag{43}$$

The most often used short time approximation is

$$\Gamma = 2c_0 \sqrt{\frac{Dt}{\pi}} , \tag{44}$$

obtained from Eq. (29) by neglecting the integral. Using a linear relationship between γ and Γ, the interfacial tension of a surfactant solution at $t \rightarrow 0$ is given then by

$$\gamma_{t \rightarrow 0} = \gamma_0 - 2nRTc_0 \sqrt{\frac{Dt}{\pi}} . \tag{45}$$

Here γ_0 is the surface tension of the solvent and n is 1 for non-ionic and 2 for ionic surfactants, respectively. The derivative of Eq. (44) with respect to $\sqrt{t}$ yields

$$\left(\frac{d\gamma}{d\sqrt{t}} \right)_{t \rightarrow 0} = -2nRTc_0 \sqrt{\frac{D}{\pi}} . \tag{46}$$

For a diffusion-controlled adsorption at a surface of constant area, the experimental data must give a straight line when plotted as $\gamma(\sqrt{t})$. The slope of this plot is equal to the expression on the right hand side of Eq. (45) and thus yields the diffusion coefficient.

Additional short and long time approximations have been summarised by Fainerman et al. (1994) based on diffusion-controlled, barrier-controlled and mixed kinetic models. An analysis of the known long time approximations was given by Makievski et al. (1997). The long time approximation derived by Joos (1999) reads

$$\Delta \gamma \big|_{t \rightarrow \infty} = \gamma(t) - \gamma_{eq} = \frac{RT\Gamma^2}{2c_0} \left(\frac{\pi}{Dt} \right)^{1/2} , \tag{47}$$

its application however requires knowledge of the adsorption isotherm so that Γ and c_0 can be inserted to obtain the diffusion coefficient D from the slope of the plot $\gamma(1/\sqrt{t})$. Recently it was shown by Daniel and Berg (2001) that the coefficient $\sqrt{\pi}$ should be replaced by $\sqrt{4/\pi}$.

4   Dynamic Surface Tension of Solutions

The surface tension of surfactant solutions is the easiest accessible experimental quantity and hence the most frequently used method to study the adsorption process at liquid interfaces. As earlier shown the rate of adsorption is a function of surface activity and bulk concentration. This explains why a broad time interval has to be experimentally covered to study the large variety of surfactants. A single method cannot provide a sufficiently broad interval so that different complementary methods are needed. Some methods are particularly developed for the short adsorption times, such as the bubble pressure method providing data from less than 1 ms up to some minutes. On the contrary, so-called static methods like the Wilhelmy plate or drop and bubble shape methods give access to very large times, starting from few seconds and reaching up to hours and even days. Both techniques complement each other perfectly.

4.1 The Maximum Bubble Pressure Technique

More than 150 years ago Simon (1851) proposed the maximum bubble pressure method for measuring the surface tension of liquids. The physical processes taking place during the growth at and separation of a bubble from the tip of a capillary, the problems of measuring bubble pressure, lifetime and dead time were considered systematically (Fainerman and Miller 1998, Kovalchuk and Dukhin 2001). Using this method, significant short time adsorption results have been obtained recently in many fields, including industrial and biological applications (Bendure 1971, Garrett and Ward 1989, Mysels 1986, 1989, Eastoe et al. 1998, Kazakov et al. 2000).

The advanced measurement technique and quantitative theoretical understanding of all sub-processes are the reason why various types of bubble pressure tensiometers are commercially available today. In some instruments, the surface tension is directly determined from the maximum pressure in the instrument's gas system connected to the capillary. The lifetime of the bubble (t_l) and the dead time (t_d) are estimated from the change in the measured pressure: t_l from the increasing part, and td from the decreasing part of the pressure curve. This procedure is self-contradictory, because precise measurements of pressure require sufficiently large system volume (Kovalchuk and Dukhin 2001, Mishchuk et al. 2001), while the precise measurement of the bubble lifetime can be performed only if the system volume is relatively small. Very short adsorption times can be reached therefore only by employing a special procedure based on the determination of a critical pressure (or critical flow rate) in the pressure - gas flow rate curve (Fainerman and Miller 1998). The point of critical flow rate represents the transition between single bubble formation and the gas jet regime, so that the time interval in this point is equal to the dead time. Recently an essentially new method for the determination of the surface lifetime was proposed, based on measurements of the dead time and lifetime from the oscillations of the gas flow from the measurement system to the capillary (Fainerman et al. 2003b).

This paragraph summarised the state of the art of the bubble pressure tensiometry and presents experimental results showing the capacity of this technique for a deeper understanding of adsorption kinetic mechanisms. The subsequent paragraph describes the drop and bubble shape methodology and demonstrates the complementarity of the two experimental techniques.

4.2 Dynamic Surface Tensions from Milliseconds to Minutes

4.1.1 General design of bubble pressure tensiometers

The design of a bubble pressure tensiometer changes from instrument to instrument, according to the specific measurement procedures. As an example, Fig. 14 illustrates the schematic diagram of the tensiometer BPA (SINTERFACE) equipped with a gas flow oscillation analyser to measure the bubble surface lifetime.

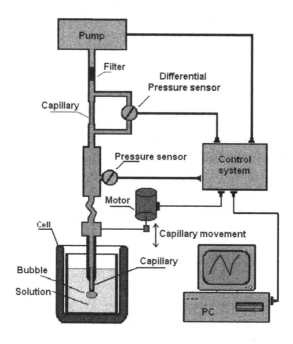

Figure 14. Schematic diagram of maximum bubble pressure tensiometer using a gas flow oscillation analyser (BPA, SINTERFACE Technologies).

The pneumatic system comprises of various parts. The air or any other gas is pressed by the compressor through the pneumatic volume, capillary and pneumatic volume, which smoothen the flow and pressure at the measuring system inlet. The air flow is determined from the pressure difference along the flow capillary via a differential pressure sensor. The control of the volume of separating bubbles is arranged by a deflector located opposite to the capillary at a definite distance (Miller et al. 1994). The optimum internal gas volume of the instrument and selection of the right capillary are two important points in the design of the device as discussed below.

The measurement procedure is typically very simple, however, measurements at very short adsorption times require special routines. In Fig. 15 an example of a dependence P(L) is shown, containing indications of the various phases of bubble growth.

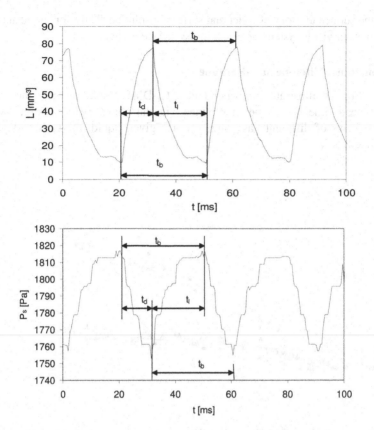

Figure 15. Dependencies L(t) (top) and P_s(t) (bottom) for water (bubble time t_b = 20 ms) obtained for a system volume V = 1.5 ml.

To calculate the surface tension γ from the simplified Laplace equation, the capillary pressure P as difference between the system pressure P_s, and the hydrostatic pressure P_h is used:

$$\gamma = \frac{r_{cap}}{2}\left(P_s - P_h\right) \qquad (48)$$

Here the deviation of the bubble from sphericity and any dynamic corrections are disregarded. For capillaries of small radii (ca. 0.1 mm) the bubbles are essentially spherical and the error in the calculation of γ from Eq. (48) is less than 0.5% and calibration with a know liquid almost eliminates this error. For wider capillaries, the error in the determined surface tension of water can be quite large and for surfactants solutions, this error increases and is roughly proportional to the ratio between the surface tensions of water and surfactant solution. Thus, for surfactant solutions a calibration with water yields systematic errors in the surface tension.

4.1.2 Determination of lifetime and deadtime

The results of Fig. 16 illustrate for water and a $C_{12}DMPO$ solution how the bubble time $t_b = t_d + t_l$ and dead time t_d can be determined via the gas flow oscillation method with measurement systems of different gas volume. For a given liquid, the t_d vs t_b dependence is almost independent of V_s.

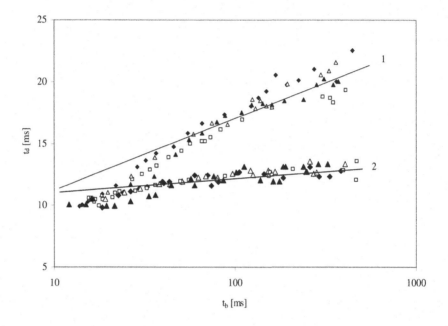

Figure 16. The dependence of the dead time td on the bubble time $t_b = t_d + t_l$ for water (2) and $C_{12}DMPO$ solution (1) for the capillary radius 0.85 mm and system volumes $V_s = 1.5$ ($\blacklozenge$), 3.7 ($\square$), 4.5 ($\triangle$) and 20.5 ml ($\blacktriangle$); lines 1 ($C_{12}DMPO$ solution) and 2 (water) are calculated from Eq. (49).

This fact is in agreement with the maximum bubble pressure theory (Fainerman and Miller 1998), and indicates that the use of a bubble deflector is important. The deflector is employed to ensure a stable bubble size irrespective of the formation frequency, and, therefore, the dead time remains approximately constant for a fixed pressure in the system. This can be seen from the data for water as shown in Fig. 16. The variation of the system pressure is taken into account via the Poiseuille equation. The lines 1 and 2 were calculated from the system pressure according to the procedure:

$$t_d = t_d^* \frac{P_s^*}{P_s} . \tag{49}$$

The asterisks refer to the corresponding parameters in the critical point. Therefore, the dead time in the critical point t_d^* (shown by an arrow in Fig. 16) for the capillary employed is 11.0 and 11.5 ms for water and the $C_{12}DMPO$ solution, respectively. The lines calculated from Eq. (49) agree quite well with the measured values. At the same time, the results indicate that the minimum dead time determined from gas flow oscillations is about 1 ms lower than the calculated value. This difference results from the hydrodynamic effects which occur at high bubble formation frequencies and are considered so far only in the software of the BPA. At a fixed bubble volume, any increase of the capillary radius results in a decrease of the dead time, and vice versa.

4.1.3 Influence of the system volume on measured dynamic surface tensions

It was mentioned above that, to eliminate the errors caused by incorrect capillary radius and bubble non-sphericity, a bubble pressure tensiometer can be calibrated with respect to water, with the known reference surface tension of 72.75 mN/m at 20°C. In Fig. 17 the dependencies of apparent dynamic surface tension of water at 20°C are shown for various volumes of the measuring system. To eliminate additional dynamic effects, the calibration was here performed in the surface lifetime range of 150 to 500 ms and at the same hydrostatic pressure (ca. 100 Pa, corresponding to a capillary immersion depth of 10 mm). One can see that in the surface lifetime range of 100 to 500 ms, the results of the measurements made for different system volumes reproduce the reference value within 0.1 mN/m. In the t_l range below 100 ms the measured (apparent) surface tension is higher than the reference value. This is due to the hydrodynamic effects caused by the rapid bubble growth and has to be accounted for by a correction factor (Fainerman and Miller 1998). It should be noted that for a small system volume ($V_s = 1.5$ cm^3) the increase in the apparent surface tension value is significantly higher. This additional effect is predicted by the theory (Kovalchuk and Dukhin 2001, Mishchuk et al. 2001) where the use of small volume reservoirs is considered.

The system volume V_s has even stronger effects on the dynamic surface tension of surfactant solutions. For a system volume $V_s = 1.5$ cm^3 the error in the measured γ values, as compared with the value for $V_s = 20.5$ cm^3, is 5 to 10%. The results of systematic studies presented elsewhere (V.B. Fainerman and R. Miller 2003) can serve as a guide for a rational choice of the measuring system volume which ensures precise measurement of the lifetime and surface tension. The optimum system volume, if all factors are taken into account, is that for which the ratio of system to bubble volume V_s/V_b is in the range between 2.000 and 5000.

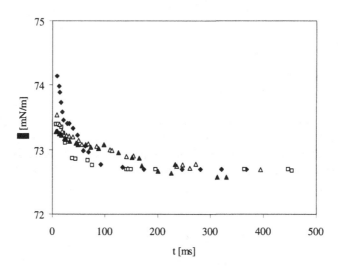

Figure 17. The dependence of apparent dynamic surface tension of water at 20°C and measuring system
volume V_s = 1.5 ($\blacklozenge$), 3.7 ($\square$), 4.5 ($\triangle$) and 20.5 ml ($\blacktriangle$).

This additional effect is predicted by the theory (Kovalchuk and Dukhin 2001, Mishchuk et al. 2001) where the use of small volume reservoirs is considered.

The system volume V_s has even stronger effects on the dynamic surface tension of surfactant solutions. For a system volume V_s = 1.5 cm^3 the error in the measured γ values, as compared with the value for V_s = 20.5 cm^3, is 5 to 10%. The results of systematic studies presented elsewhere (V.B. Fainerman and R. Miller 2003) can serve as a guide for a rational choice of the measuring system volume which ensures precise measurement of the lifetime and surface tension. The optimum system volume, if all factors are taken into account, is that for which the ratio of system to bubble volume V_s/V_b is in the range between 2.000 and 5000.

4.1.4 Determination of the hydrodynamic pressure

The pressure measured in a maximum bubble pressure instrument includes the hydrostatic pressure, as shown by Eq. (48). The immersion depth of the capillary h has to be measured therefore accurately, except the capillary tip is at the liquid level (Brown 1932) or if two capillaries are used, as proposed by Sugden (1922). Due to our opinion, set-ups with two capillaries can be successfully used only in studies of pure liquids or highly concentrated solutions. However, for usual surfactant solutions, due to the dynamic character of the surface tension, significant errors can arise, caused by the fact that surface tension in the bubbles of different volume is different.

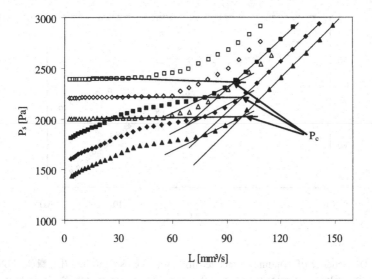

Figure 18. Pressure in the measuring system as a function of the flow rate L for water ($\square$,$\diamond$,$\triangle$) and C$_{12}$DMPO solution ($\blacksquare$, $\blacklozenge$, $\blacktriangle$) for different hydrostatic pressure values 100 Pa ($\triangle$, $\blacktriangle$), 300 Pa ($\diamond$, $\blacklozenge$) and 500 Pa ($\square$, $\blacksquare$).

A rough estimation of the hydrostatic pressure can be performed with a single large capillary if a low V_s/V_b ratio is used (Patent 1997, 1999). In this case the minimum in the $P_s(t)$ curve is approximately equal to the hydrostatic pressure. In other tensiometers the critical point in the $P_s(L)$ dependence can be used for the estimation of the hydrostatic pressure. Examples of such dependencies for water and a C$_{12}$DMPO solution, obtained for various values of the hydrostatic pressure (100 to 500 Pa) are shown in Fig. 18.

The critical pressure P_c shown by the arrows for each given hydrostatic pressure in the case of C$_{12}$DMPO solution is exactly equal to the pressure measured for water $P < P_c$ at the same hydrostatic pressure. The advantages of such a procedure are discussed in detail elsewhere (Fainerman et al. 2003b). However, to perform precise measurements of dynamic surface tension, one should measure the hydrostatic pressure accurately. This can be done by a direct determination of the immersion depth of the capillary as designed in the BPA tensiometers of SINTERFACE. The larger the capillary radius is the higher must be the accuracy of the immersion depth.

4.1.5 Measurement of dynamic surface tensions at longer adsorption times

Actually, the maximum bubble pressure technique is not suitable for long time experiments. Due to the small size of a single bubble, the experiment is very sensitive to smallest temperature changes. However, for relatively large system volumes, a self-generation of bubbles with longer lifetimes is possible. This method as known as the so-called stopped flow method (Fainerman and Miller 1998).

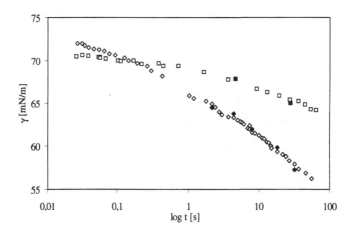

Figure 19. Dependence of dynamic surface tension for cerebrospinal fluid ($\square$, $\blacksquare$) and blood serum ($\diamond$, $\blacklozenge$) using the Peltier generator ($\square$, $\diamond$) and the stopped flow regime ($\blacklozenge$, $\blacksquare$) (Kazakov et al. 2000).

To extend the time interval for bubble pressure tensiometers, special gas supply systems are required, because usual gas pumps cannot be accurately controlled at very low flow rates. One of the efficient possibilities is the use of a Peltier bubble generator with which bubble times of up to 100 s can be reached (installed in the BPA instruments, Fainerman et al. 2003a). The generator can be connected to the gas volume (see Fig. 14) and controlled by the computer. The compressor can be cut off from the measuring system by a valve while in the same moment the Peltier generator is activated. The design of the generator with an internal volume of about 3 cm^3 allows generation of about 100 bubbles via heating of the Peltier elements to a maximum temperature. For slow surface tension decrease a rapid heating of the Peltier elements is advantageous, while for faster surface tension decrease the slower heating regime is useful. Examples obtained in studies of biologic liquids are shown in Figure 19 together with data obtained by the stopped flow procedure.

4.1.6 Effect of the capillary geometry on the dead time

The dead time t_d in the critical point can be calculated from $t_d = V_b/L_c$ and for any system pressure P and gas flow L, a correction to the dead time can be introduced via Eq. (49). The critical flow, in turn, is determined by the capillary length and diameter. For a fixed capillary radius one can control the critical flow rate L_c, and therefore the dead time, by varying the capillary length. For a wide capillary, for example with a diameter $2r_{cap} \geq 0.25$ mm, the respective capillary length can be very long, so that this procedure is not suitable for the control of the critical flow. This obstacle can be overcome by using combined narrow and wide capillaries, arranged by a wide capillary immersed into the liquid and connected to a narrow capillary, or by decreasing the cross-section of a wide capillary in its upper part by a special insert (Fainerman et al. 2003b). Note that these methods could be used to control the dead time only in systems for which the condition $V_b \ll V_s(P_s/P_a)$ is fulfilled. This condition corresponds to the situation when the gas mass excess in the system is large enough to enable a spontaneous formation of bubbles during the dead time interval. From the maximum bubble pressure theory the critical gas flow L_c can be readily calculated from the aerodynamic resistance of the capillary via the following relationship for a capillary of arbitrary geometry (Fainerman and Miller 1998)

$$\frac{P_c}{L_c k_P} = const = K \,, \tag{50}$$

where $k_P = 8\eta l / \pi r_{cap}^4$ is the Poiseuille law coefficient and depends on the gas viscosity η and the capillary length l and radius r_{cap}. A comprehensive study was published recently by Fainerman et al. (2003b) showing the possibilities of deadtime control via the geometry of the capillary.

4.2 Dynamic Surface Tensions from Milliseconds to Minutes

At liquid interfaces most surfactants adsorb diffusion controlled. Over many years various models were derived and the nature of barriers in adsorption or desorption processes analysed. In the eighties, however, it was understood by Lunkenheimer and Miller (1987, 1988) from a systematic analysis of impurity effects that in most cases surfactants adsorb according to a diffusion kinetics. The possible barriers discovered in literature were mimicked by impurities, traces of highly surface active compounds.

Apart from impurities, the formation of surfactant aggregates at the interface makes the adsorption process slower than expected from diffusion. Applying the classical diffusion model to such data would result in a diffusion coefficient significantly smaller than physically reasonable and hence a barrier mechanism would have to be postulated. It will be shown below that a correct consideration of surface aggregation processes allow to quantitatively describe such systems in the framework of a purely diffusion controlled mechanism.

In contrast to surface aggregation, changes in the molar area of adsorbed molecules can lead to an apparently enhanced adsorption process. Such a so-called "super-diffusion" process can be explained by considering changes in the molar surface area with changing surface coverage, as given by the reorientation model of Eqs. (18) to (20). Such surfactant systems are then again quantitatively understood by a purely diffusion controlled model. Examples are given below.

4.2.1 Diffusion controlled adsorption of simple surfactants

For various non-ionic surfactants the dynamic surface tension data can be interpreted in terms of a diffusion controlled adsorption mechanism easily (Fainerman and Miller 1995). Some selected experimental results for different Triton solutions ($C_{14}H_{20}O\,(C_2H_4O)_m\,H$) with different numbers m of ethylene oxide groups are shown as a $\gamma(\sqrt{t})$ -plot. For all solutions we get a surface tension $\gamma(t) = \gamma_0$ at t=0 which is the condition for the application of Eq. (46).

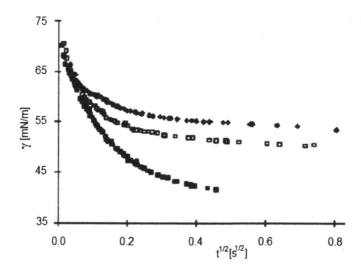

Figure 20. Dynamic surface tensions of Triton X-m solutions as function of √t; m=16.5, $c_o = 1.07.10^{-6}$ mol/cm³ (◆); m=30.5, $c_o = 4.52.10^{-7}$ mol/cm³ ( ); m=40.5, $c_o = 5.08.10^{-7}$ mol/cm³ (■)

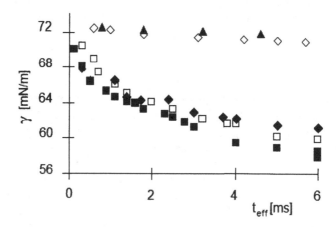

Figure 21. Dynamic surface tensions of different aqueous Triton X-m solutions plotted as function of the effective surface age t_{eff}; m=10.0, $c_o = 1.55.10^{-7}$ mol/cm³ (▲); m = 8, $c_o = 4.23.10^{-7}$ mol/cm³ (◇); m=16.5, $c_o=1.07.10^{-6}$ mol/cm³ (◆); m=30.5, $c_o=4.52.10^{-7}$ mol/cm³ ( ); m=40.5, $c_o=5.08.10^{-7}$ mol/cm³ (■)

For non-ionic surfactants, in Eq. (46) we have n = 1, and the slope of the linear part of the curves $\gamma(\sqrt{t})$ give direct access to the diffusion coefficient D. Reasonable values for D would then proof that the surfactant adsorbs diffusion controlled. The slopes for the three solutions are similar and of the order of 200 mN/(m s$^{-1/2}$), which yields diffusion coefficients of the order of D=10^{-4} cm²/s. These values are obviously by more than one order of magnitude too large, however, the Tritons are technical surfactants and therefore a quantitative understanding will be impossible.

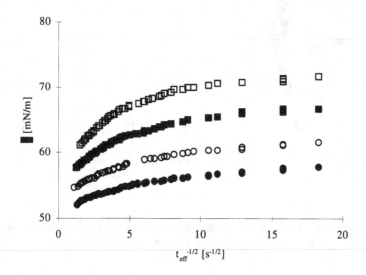

Figure 22. Dynamic surface tension γ of Triton X-405 as a function of 1/√t at different concentrations, co = 1.53 10^{-7} ( ), 2.54 10^{-7} (■), 5.08 10^{-7} (O), 1.016 10^{-6} (●) mol/cm³

In case the experimental curves do not extrapolate to a surface tension of pure water at t =0, an initial load of the bubble surface has to be considered. A quantitative analysis of this situation was elaborated by MacLeod and Radke (1994) in which the initial state of adsorption at the bubble surface was considered.

The data shown in Fig. 21 are partially those of Fig. 20, however, plotted as a function of the effective surface age and restricted to the short time interval t < 5 ms. This demonstrate that the maximum bubble pressure technique is able to provide results in the sub-millisecond time range.

The dynamic surface tensions of Triton X-m solutions at longer adsorption times were measured systematically by Fainerman and Miller (1995a). As an example results for two of the Tritons at different concentrations are shown in Figs. 22 and 23. The surface activity of the Triton compounds differs and hence the dynamic surface tensions have a different shape. While the Tritons with a large number of EO groups are less surface active and produce a surface tension change only at higher concentrations (Triton X-405 with statistically 40.5 EO units), those with the shorter EO chains adsorb stronger (Triton X-100 with 10 EO units).

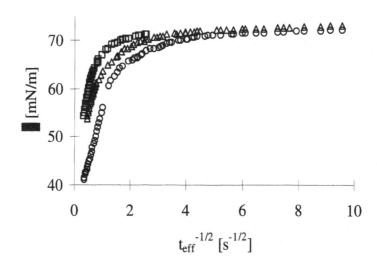

Figure 23. Dynamic surface tension γ of Triton X-100 as a function of $1/\sqrt{t}$ at different concentrations, $c_0 = 7.1 \cdot 10^{-8}$ ( ), $1.24 \cdot 10^{-7}$ (Δ), $1.55 \cdot 10^{-7}$ (O) mol/cm³

When the dynamic surface tensions are plotted as functions $\gamma(1/\sqrt{t})$ we can expect linear dependencies for $t \to \infty$, i.e. $1/\sqrt{t} \to 0$, according to Eq. (47). The data analysis based on this equation confirms that the Tritons adsorb diffusion controlled.

4.2.2 Surfactants able to change their orientation

From time to time it is observed in literature that adsorption processes proceed faster than expected from a diffusion mechanism. This can of course be ascribed to convection in the bulk, however, this is not a good explanation for surfactant systems, where such behaviour is observed under various experimental conditions.

As mentioned above reorientation processes in the adsorption layer can mimic adsorption processes faster than expected from diffusion. For modelling the adsorption process with molecular reorientation the Ward and Tordai equation (37) as the most general relationship between the dynamic adsorption $\Gamma(t)$ and the subsurface concentration $c(0,t)$ can be used. As additional relationship the reorientation isotherm Eqs. (18) to (20) are included into the model.

The adsorption dynamics of Tritons have been described systematically in the monograph by Fainerman et al. (2001) and a diffusion controlled kinetics based on a Langmuir isotherm was suggested, also from the data given above. It was shown in the same monograph, that oxethylated surfactants can be better described by a reorientation isotherm (Fainerman et al. 2000a). This is however only true for high quality samples such as define compounds of the type C_nEO_m. An analysis of dynamic surface and interfacial tensions for example of the surfactant $C_{10}EO_8$ solutions was given by Miller et al. (1999). Selected dynamic surface tensions for $C_{10}EO_8$ at the water/air interface, as measured by Chang et al. (1998) by the pendent bubble method, are shown in Fig. 24 and compared with calculations for the Langmuir and the reorientation models.

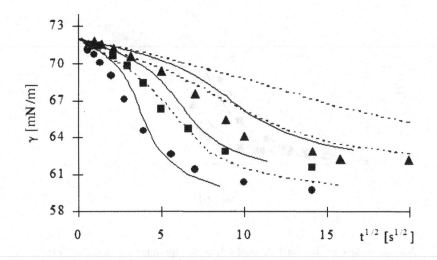

Figure 24. Dynamic surface tension for $C_{10}EO_8$ solutions at three concentrations: $c_0 = 4\cdot10^{-9}$ mol/cm³ (▲), $c = 6\cdot10^{-9}$ mol/cm³ (■), $c = 10^{-8}$ mol/cm³ (●), data from Chang et al. (1998), $D = 5\ 10^{-6}$ cm²/s, Langmuir model (dotted lines), reorientation model (solid lines).

One can see that the Langmuir model overestimates the dynamics, i.e. the value of D required to match the experimental data would be 3 times higher than the physically expected value, i.e. 1.5·10-5 cm²/s. At the same time, the reorientation model agrees well with the experimental data when a diffusion coefficient of D = 5 10-6 cm²/s is used. Further experimental examples and a detailed analysis of a non-instantaneous reorientation process on the adsorption kinetics were given by Miller et al. (1999).

4.2.3 Surfactants undergoing 2D-aggregation

As shown in the previous example, interfacial reorientations can explain adsorption processes which appear as if they are faster than diffusion. By using the correct adsorption isotherm the diffusion mechanism yields perfect agreement to the experimental data with a normal diffusion coefficient.

The variety of explanations for a slower surface tension decrease, a case which is often experimentally observed, is very large and was summarised by Fainerman et al. (2001). One of the reasons for slower adsorption kinetics can be an inaccurate adsorption isotherm. As experimental example, the dynamic surface tensions for 1-decanol solutions by Lin et al. (1994) are compared in Fig. 25 with the results calculated from the aggregation model.

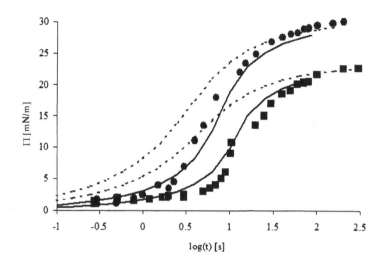

Figure 25. Dynamic surface tension for 1-decanol solutions, experimental results from Lin et al. (1994) for $c_o = 6.32 \cdot 10^{-5}$ mol/l ($\bullet$) and $c_o = 1.018 \cdot 10^{-4}$ mol/l ($\blacksquare$); dotted lines - Langmuir model, solid lines - aggregation model of Eqs. (21), (22) with n = 2.5 and $\Gamma_c \leq 10^{-9}$ mol/m²

In the calculations of $\gamma(t)$ the following parameters for the decanol adsorption isotherm at the solution/air interface were used: $\omega_1 = 1.67 \ 10^9$ cm²/mol, a = 1.8 and b = $1.21 \cdot 10^7$ cm³/mol. One can see that the aggregation model provides satisfactory agreement with experimental data, while the curves calculated from the Langmuir isotherm differ significantly within the short time range. For an aggregation number n < 2 or n > 3, the experimental data deviate much from the theory, which strongly supports the idea that decanol molecules form dimers and trimers in the adsorption layer.

To summarise, more experimental data have to be produced to demonstrate the suitability of the aggregation model. However, the given example shows impressively, that this model can describe experimental data very well without the assumption of an adsorption barrier, i.e. within the framework of the pure diffusion- controlled adsorption.

5   Drop and Bubble Shape Experiments

The method with the largest capacity is obviously the drop and bubble shape analysis, in the literature often named ADSA. This methodology does not provide data at extremely short adsorption times, however, it has quite a number of advantages. First of all it is applicable to liquid/gas and liquid/liquid interfaces, it requires very small amounts of samples, and it is easy to temperature control. Moreover, it gives access to surface rheology. Equipped with an additional pressure sensor, it allows even measurements at adsorptions times of less than 0.1 s. Also studies at interfaces between two liquids of equal density are possible, where all the other methods fail.

5.1 Drop and Bubble Profile Analysis Method

The presently most powerful technique for obtaining the liquid-vapour or liquid-liquid interfacial tension is based on the shape of a drop or bubble. In essence, the shape of a drop or bubble is determined by balance of surface tension and gravity effects. Surface forces tend to make drops spherical whereas gravity tends to elongate a pendant drop or buoyant bubble. Fig. 26 shows the schematic of an experimental set-up (for details see Chen et al. 1998, Loglio et al. 2001).

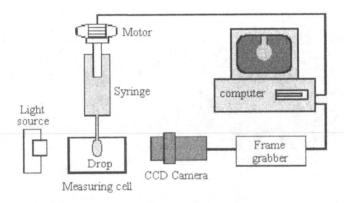

Figure 26. Schematic of a drop or bubble shape analysis set-up

This profile analysis technique has numerous advantages: only small amounts of the liquid are required, it is suitable for both liquid-vapour and liquid-liquid interfaces, it is applicable to materials ranging from organic liquids to molten metals and from pure solvents to concentrated solutions, it has no limitation to the magnitude of surface or interfacial tension, and is user-friendly in a broad range of temperatures and pressures. The available time interval reaches from part of a second up to hours and even days so that extremely slow processes can be easily followed.

To describe a liquid meniscus and hence to obtain the interfacial tension from the profile coordinates the Gauss - Laplace equation is used. This equation represents the mechanical equilibrium for two homogeneous fluids separated by an interface (Neumann and Spelt 1996). It relates the pressure difference across a curved interface to the surface tension and the curvature of the interface

$$\gamma\left(\frac{1}{R_1} + \frac{1}{R_2}\right) = \Delta P, \tag{51}$$

where R_1 and R_2 are the two principal radii of curvature, and ΔP is the pressure difference across the interface. In the absence of any external forces other than gravity g, ΔP is a linear function of the elevation z:

$$\Delta P = \Delta P_0 + (\Delta\rho)gz .$$

(52)

ΔP_0 is the pressure difference at a reference plane, $\Delta\rho$ is the density difference between the two bulk phases, g is the gravitational acceleration, and z is the vertical height of the drop measured from the reference plane, the drop/bubble apex.

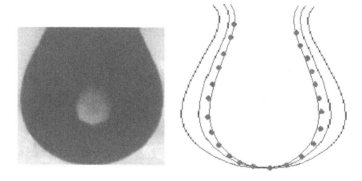

Figure 27. Video mage of a drop and schematic of fitting the Laplace equation to shape coordinates

The earliest efforts in the analysis of axisymmetric drops were those of Bashforth and Adams (1883) who prepared tables of characteristic coordinates. Later Hartland and Hartley (1976) used a computer program to integrate the appropriate form of the Laplace equation. The results were again presented in tables. A significant step ahead was done by the group of Neumann through the development of the software package called ADSA - Axisymmetric Drop Shape Analysis. This was the first commercial software for fitting the Laplace equation to the shape of drops or bubbles obtained as a video image (Rotenberg et al 1983, Cheng et al. 1990, 1998, Neumann and Spelt 1996). Today quite a number of commercial set-ups and respective software packages exist yielding essentially the same quality of results. However, the comfort of handling and the number of available features of the various instruments are very different.

In recent years, several theoretical and experimental attempts have been performed to develop methods based on oscillations of supported drops or bubbles. For example, Tian et al. (1997) used quadrupole shape oscillations in order to estimate the equilibrium surface tension, Gibbs elasticity, and surface dilational viscosity. Pratt and Thoraval (1997) used a pulsed drop rheometer for measurements of the interfacial tension relaxation process of some oil soluble surfactants. The oscillating bubble system uses oscillations of a bubble formed at the tip of a capillary. The amplitudes of the bubble area and pressure oscillations are measured to determine the dilational elasticity while the frequency dependence of the phase shift yields the exchange of matter mechanism at the bubble surface (Kretzschmar and Lunkenheimer 1970, Wantke and Fruhner 1998). A comprehensive analysis of oscillating drops and bubbles has been performed recently and new mechanisms proposed to analyse the data quantitatively (Kovalchuk et al. 2003).

The instrument shown schematically in Fig. 26 is suitable for slow oscillation experiments, as it was performed for the first time by Miller et al. in 1993. The frequency limit of the oscillations is given by the condition for the liquid meniscus shape, which has to be Laplacian. Under too fast deformations this condition is not fulfilled and hence the method does not provide reliable results. To reach higher frequencies of oscillation, the above mentioned oscillating drop or bubble experiments are suitable, because the shape of the menisci is spherical due to the small diameter. The instrument of Fig. 26 can be designed such that a pressure sensor and piezo translator are built in and the video system serves as optical control and determines the drop/bubble diameter accurately.

For very small sinusoidal changes of the interfacial area of spherical drops or bubbles the pressure amplitude can be calculated from the force balance. The measured pressure amplitude can be written as the sum of a geometric (radius) component, a contribution caused by changes in the interfacial tension $\Delta\gamma$, and a hydrodynamic contribution. The dynamic pressure can be described by Eq. (4.129):

$$\Delta P = \frac{2\gamma\Delta r}{(r_0)^2} - \frac{2\Delta\gamma}{r_0} + \Delta\tilde{P}_{hydrod.} \tag{53}$$

with $r_0 = (r_1 + r_2)/2$. The first term is a geometric component only controlled by the capillary size and the interfacial tension of the pure system. The second term is used to calculate the dilational properties. $\Delta\gamma$ as the change in interfacial tension also incorporates the elastic and viscous contributions. The third term in Eq. (53) is controlled by hydrodynamic effects, such as inertia and viscosity effects.

5.2 Dynamic Surface Tensions from Seconds to Days

The mass of data provided by the profile analysis tensiometry becomes clear from Fig. 28, where the dynamic surface tensions of 9 concentrations of a model surfactant $10_{10}EO_4$ are shown. As mentioned above, the PAT (SINTERFACE Technologies) yields data in the time range from few seconds up to hours or even days. While the curve at the lowest concentration (10^{-5} mol/l) starts at a value close to the surface tension of water, the curves for the higher concentrations begin at much lower values. This is clear from the adsorption kinetics: at a surface age of say 5 s a significant amount of surfactants has already been adsorbed. At the highest concentration of $7 \cdot 10^{-4}$ mol/l there is even no kinetics measurable over the entire time range.

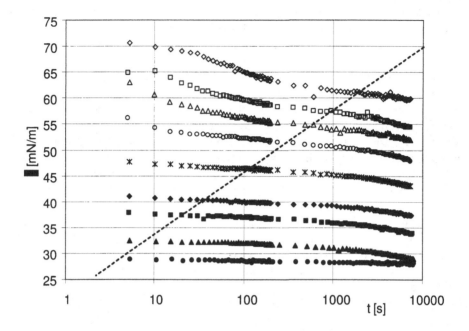

Figure 28. Dynamic surface tensions of $C_{10}EO_4$ solutions at various concentrations c: 10^{-5} ($\diamond$), $2\ 10^{-5}$ ($\square$), $3\ 10^{-5}$ ($\triangle$), $5\ \ 10^{-5}$ ($\bigcirc$), 10^{-4} ($\times$), $2\ 10^{-4}$ ($\blacklozenge$), $3\ 10^{-4}$ ($\blacksquare$), $5\ 10^{-4}$ ($\blacktriangle$), $7\ 10^{-4}$ ($\bullet$) mol/l; dotted line – approximate time for surface tensions close to equilibrium

When we use the respective plot $\gamma(1/\sqrt{t})$ a reasonable values of the equilibrium surface tension can be estimated. The same data of Fig. 28 are shown in this $1/\sqrt{t}$-plot are shown in Fig. 29. The linear extrapolation to $t \to \infty$ gives the isotherm shown in Fig. 30 by the symbols ($\diamond$). When we however analyse the data given in Fig. 28 we can see that after the establishment of a temporary equilibrium the surface tension starts again to decrease. This phenomenon is typical for surfactant mixtures, and will be discussed in more detail at the end of this chapter. In order to find the properties of the main surfactant, in the present case of $C_{10}EO_4$, we have to analyse the data of Fig. 28 in a different way. The dotted line in this figure marks the time for each concentration, where the equilibrium surface tension should have been reached in case diffusion controls the adsorption process. Using these values, another surface tension isotherm is obtained, given in Fig. 30 by the symbols ($\square$).

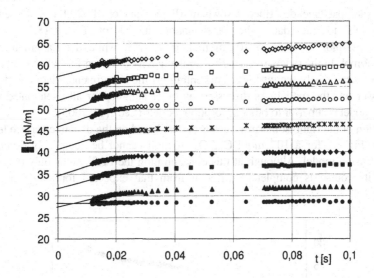

Figure 29. Dynamic surface tensions of $C_{10}EO_4$ solutions at various concentrations (same as in Fig. 28), solid lines – extrapolation to estimate the equilibrium surface tensions

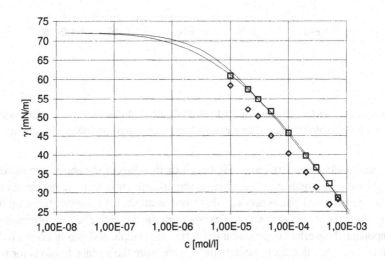

Figure 30. Surface tension isotherm of $C_{10}EO_4$ at the water/air interface; $\diamond$ - extrapolated values $\gamma(t)|_{t\to\infty}$, $\square$ - values $\gamma(t)$ at a time needed to reach equilibrium according to a diffusion model (see dotted line in Fig. 28), solid line – reorientation model with $\omega_1 = 6.7 \; 10^5$ m²/mol, $\omega_2 = 2.56 \; 10^5$ m²/mol, $\alpha = 1.18$; dotted line – Langmuir isotherm with $\omega = 2.9 \; 10^5$ m²/mol

The software package described elsewhere (Fainerman et al. 2001) was used to fit two isotherm models to the data: the reorientation model of Eqs. (18) to (20) yields $\omega_1 = 6.7 \ 10^5$ m²/mol, $\omega_2 = 2.56 \ 10^5$ m²/mol, $\alpha = 1.18$; while the Langmuir model gives $\omega = 2.9 \ 10^5$ m²/mol. In the range given in Fig. 30 the quality of fitting for both models is the same, however, as one can easily see, at lower concentrations the experimental data could give a more accurate answer on the model. As discussed by other authors, for all oxethylated alcohols the reorientation model is superior to other adsorption models (Lee et al. 2003).

A discussion of the right isotherm values is easiest based on the dynamic surface tensions. The data given in Fig. 31 for a 10^{-4} mol/l $C_{10}EO_4$ solution cannot be adequately described by the respective diffusion model. Although at this concentration the PAT yields only data close to the equilibrium, its course is completely different from that expected for a diffusion controlled process.

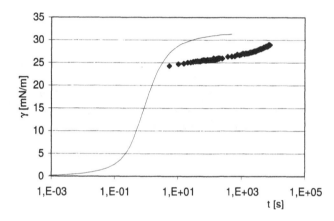

Figure 31. Dynamic surface tensions of a 10^{-4} mol/l $C_{10}EO_4$ solution, solid line – diffusion model ($D = 5 \ 10^{-6}$ cm²/s) using the final extrapolated surface tension values as isotherm

When we use the reorientation isotherm as obtained on the basis of the above described algorithm (surface tension values at times corresponding to the adsorption times), we get the picture shown in Fig. 32. The experimental points are well described at all shown concentrations, up to a certain time moment, where the experimental data deviate from the theoretical curve, obviously due to an impurity component. This effect depends on the surfactant sample and was not observed by Lee et al. (2003). Note, however, that these authors did not measure the surface tensions for times larger than 1 hour so that the deviation was maybe not remarkable enough to be detected.

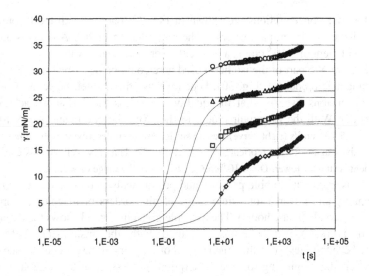

Figure 32. Dynamic surface tensions of some $C_{10}EO_4$ solution, concentrations: $2 \ 10^{-5}$ mol/l ($\Diamond$), $5 \ 10^{-5}$ mol/l ($\square$), 10^{-4} mol/l ($\triangle$), $2 \ 10^{-4}$ mol/l ($\bigcirc$), solid lines – reorientation model ($D = 5 \ 10^{-6}$ cm²/s) using the correct equilibrium surface tension values as isotherm

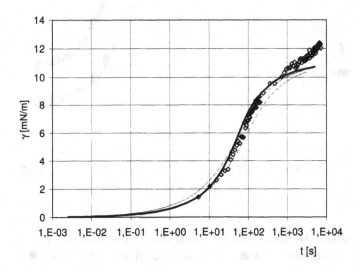

Figure 33. Dynamic surface tensions of a 10^{-5} mol/l $C_{10}EO_4$ solution ($\Diamond$), solid line – reorientation model ($D = 4 \ 10^{-6}$ cm²/s), dotted lines – Langmuir model with $D = 8 \ 10^{-6}$ cm²/s (left curve) and $D = 5 \ 10^{-6}$ cm²/s (right curve)

The results for four concentrations are shown in Fig. 32 together with the theoretical curved calculated with the above given parameters of the reorientation model. All curves are calculated for a diffusion coefficient of $5 \cdot 10^{-6}$ cm²/s, a physically very reasonable value.

As discussed above, the part of the isotherm used here could be described with the same quality by using the Langmuir model (cf. Fig. 30). When we use this model, however, to describe the dynamic surface tensions, we see that the quality is far less than what we achieve with the reorientation model. With the Langmuir model neither with the coefficient $D = 8 \cdot 10^{-6}$ cm²/s (left curve) nor with $D = 5 \cdot 10^{-6}$ cm²/s (right curve) we can sufficiently well describe the experimental data, while the solid line, obtained again for the reorientation model, fits the data very well. At this concentration the diffusion coefficient is slightly lower, $D = 4 \cdot 10^{-6}$ cm²/s, which is in the range of accuracy.

The two methods maximum bubble pressure and profile analysis tensiometry complement each other experimentally and cover a total time range of nine orders of magnitude from about 10^4 seconds up to 10^5 seconds (many hours). The example given in Fig. 33 shows the dynamic surface tension of two Triton X-100 solutions measured with the instruments BPA and PAT (SINTERFACE Technologies) over the time interval of 7 orders of magnitude. As one can see, the experiments cover the beginning of the adsorption process and the establishment of the equilibrium state.

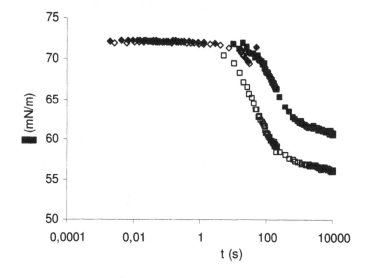

Figure 34. Dynamic surface tension of two Triton X-100 solutions:
co = $5 \cdot 10^{-9}$ mol/cm³ (◆■), 10^{-8} mol/cm³ (◇□), BPA (◆◇), PAT (■□)

5.3 Surface Dilational Elasticity of Surfactant Layers

Interfacial relaxation methods are typically based on a perturbation of the equilibrium state of an interface by small changes of the interfacial area. The ratio of the amplitudes of surface tension and relative area changes gives the modulus of elasticity ε, defined as

$$\varepsilon = -\frac{d\gamma}{d\ln\Gamma} = \frac{d\gamma}{d\ln A} = -\frac{d\gamma}{d\ln\Gamma}\frac{d\ln\Gamma}{d\ln A} \approx \frac{\Delta\gamma}{\Delta A}A_0 \tag{54}$$

There are transient and harmonic perturbations of the interfacial area, for which, as it was shown by Loglio et al. (1994), the theoretical basis is the same. For harmonic relaxation processes there is a phase difference between the generation of the oscillation and the response function which is an easy accessible measure of the exchange of matter.

As mentioned above the oscillating drop or bubble method, based on profile analysis tensiometry, is the most recently developed method to investigate the surface relaxation of soluble adsorption layers. By increasing/decreasing the volume of a pendent drop or bubble, a variety of area changes can be performed, such as step, square pulse, ramp type, trapezoidal, and of course harmonic area changes at low frequencies.

Assuming a diffusional flux to and from the interface caused by the induced area changes Lucassen and van den Tempel (1972) derived the following relationship for the frequency dependent elasticity

$$\varepsilon(i\omega) = \varepsilon_0 \frac{\sqrt{i\omega}}{\sqrt{i\omega} + \sqrt{2\omega_0}} \tag{55}$$

with

$$\varepsilon_0 = -\left(\frac{d\gamma}{d\ln\Gamma}\right)_A \quad \text{and} \quad \omega_0 = \left(\frac{dc}{d\Gamma}\right)^2 \frac{D}{2}. \tag{56}$$

Here ω is the oscillation frequency, and the parameter ω_0 is the characteristic frequency, which is inverse proportinal to the diffusion relaxation time τ_D given in Eq. (35). This characteristic frequency exists also for any transient relaxation processes. The interfacial response functions for a number of transient relaxations were discussed recently by Loglio et al. (2001). Among these, the trapezoidal area change is the most general perturbation which contains area changes such as the step or ramp type and the square pulse as particular cases.

In Fig. 35 an oscillation experiment with a Triton X-100 solution is shown with a 400 s oscillation period. This oscillation can be quantitatively interpreted by a Fourier analysis, as it is implemented in most of the profile analysis tensiometer software. Another way of data interpretation is to plot the change in surface tension over the corresponding area change. As one can see, an ellipse with a certain tilt angle and thickness results. These two geometrical properties carry the rheological information: the tilt is a measure of the dilational elasticity, and the thickness is proportional to the exchange of matter rate, also called dilational viscosity. Essentially this parameter corresponds to the phase shift between the generated area oscillation and the surface tension response. With increasing frequencies the thickness decreases while the tilt angle increases up to a final value which corresponds to the dilational elasticity modulus.

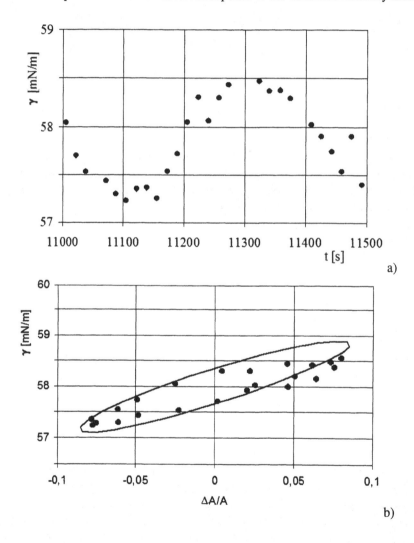

Figure 35. Surface tension response for a Triton X-100 solution at 10^{-5}mol/l; a) change of surface tension with time; b) same as in (a) but plotted over the relative area change A(t)/A during oscillation

It was discussed quite extensively, that interfacial dynamics and rheology are key properties of liquid disperse systems, such as foams and emulsions. The stability of such systems depends for example on the dilational elasticity and viscosity, however, surely not on the elasticity modulus (Borwankar et al. 1992). Here, the interfacial rheology with its frequency dependence comes into play, and data at respective frequencies will possibly correlate with the stability behaviour.

Application of tensiometry and surface rheology in medical research was shown by Kazakov et al. (2000). For example, selected dynamic surface tension values of serum or urine correlate with the health state of patients suffering from various diseases. In the course of a medical treatment these values then change from a pathological level back to the normal values determined as standard for a certain group of people (age and sex). Fig. 36 shows the surface tension response after a step-type area change of a pendant drop area by about 10% for 6 serum samples from the same patient at different stages of his acute kidney insufficiency. It appears to be efficient to use such easy to handle methods for therapy control in the medical practice.

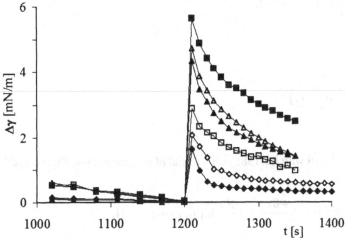

Figure 36. Surface tension response after step type perturbation of a serum sample from a 46 years old patient with acute kidney insufficiency; arrival at hospital ($\Diamond$), after therapy ($\blacklozenge$, $\square$), after haemodialysis ($\blacksquare$), polyuria ($\triangle$), leaving the hospital ($\blacktriangle$)

6 Adsorption Behaviour of Mixed Surfactant Systems

Only in fundamental studies the adsorption of pure model surfactants and proteins can be investigated. Under practical conditions the adsorbing species are typically mixtures of components of different surface activity. Often mixtures are even used on purpose to reach special effects. This makes a deeper understanding of the adsorption process from solutions of surfactant mixtures important. A rather simple model for the adsorption of mixed surfactants at a liquid interface is presented here and its capacity is demonstrated on the basis of surface tension data for model mixtures. Also the adsorption dynamics requires generalised models to quantitatively understand the process for such mixed systems. Examples for mixed non-ionic surfactant systems are discussed here.

6.1 Thermodynamics of mixed adsorption layers

There are different models developed in the past to describe the adsorption from mixed surfactant solutions, for example recently by Siddiqui and Franses (1997), Ariel et al. (1999), Mulqueen and Blankschtein (1999, 2000), Penfold et al. (2003). The simplest model is obviously a generalised Langmuir isotherm (for ideal behaviour in the bulk and at the interface) for mixtures of two surfactants 1 and 2 with similar partial molar surface area ω can be presented in the form (Fainerman et al. 2001)

$$\Pi = -\frac{RT}{\omega}\ln(1-\theta_1-\theta_2),$$

(57)

$$b_i c_i = \frac{\theta_i}{(1-\theta_1-\theta_2)}.$$

(58)

Equation (57) combined with Eq. (58) yields a kind of generalised von Szyszkowski equation

$$\Pi = \frac{RT}{\omega}\ln\left(1+\frac{\theta_1+\theta_2}{1-\theta_1-\theta_2}\right) = \frac{RT}{\omega}\ln(b_1c_1+b_2c_2+1).$$

(59)

Together with the corresponding equations for the individual surfactants 1 and 2 (the subscript 0 refers to the value for the individual solution),

$$\Pi_i = -\frac{RT}{\omega}\ln(1-\theta_{0i}) = \frac{RT}{\omega}\ln(1+b_ic_i),$$

(60)

$$b_i c_i = \frac{\theta_{01}}{1-\theta_{0i}},$$

(61)

Eq. (59) yields (Fainerman et al. 2002a)

$$\exp \overline{\Pi} = \exp \overline{\Pi}_1 + \exp \overline{\Pi}_2 - 1 .$$ (62)

Here $\overline{\Pi} = \Pi \omega / RT$, $\overline{\Pi}_1 = \Pi_1 \omega / RT$ and $\overline{\Pi}_2 = \Pi_2 \omega / RT$ are the dimensionless surface pressures of the mixture and individual solutions at the same surfactant concentrations as in the mixture, respectively.

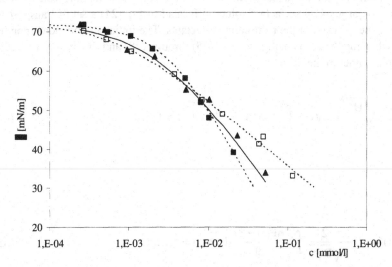

Figure 37. Surface tension isotherms for Triton X-100 ($\square$) and $C_{12}EO_5$ ($\blacksquare$), and a 1:1 mixture of the two surfactants ($\blacktriangle$), dotted lines - calculated from Frumkin's Eq. (11) for Triton X-100 ($\omega = 3.4 \cdot 10^5$ m²/mol, a = 0) and for $C_{12}EO_5$ ($\omega = 1.5 \cdot 10^5$ m²/mol, a = 0.6); solid line – calculated from Eqs. (62) and (63)

The Eq. (62) allows calculating the surface pressure of a mixture from the experimental surface tensions of the individual solutions, without any additional knowledge on the surface tension isotherms for the individual solutions. It was shown by Fainerman and Miller (2001) that this equation even holds for situations where the assumption of equal values for ω is not fulfilled. Then the dimensionless pressures are defined by $\overline{\Pi}_1 = \Pi_1 \omega_1 / RT$ and $\overline{\Pi}_2 = \Pi_2 \omega_2 / RT$, while the mean molar area is given by

$$\omega = \frac{\omega_1 \Pi_1 + \omega_2 \Pi_2}{\Pi_1 + \Pi_2} = \omega_1 \frac{\overline{\Pi}_1 + \overline{\Pi}_2}{\overline{\Pi}_1 + \overline{\Pi}_2 (\omega_1 / \omega_2)} .$$ (63)

As example, the mixture of Triton X-100 and $C_{12}EO_5$ studied by Siddiqui and Franses in 1996 can be considered. These two surfactants have very different ω_i values. Figure 37 illustrates the surface tension isotherms for the individual solutions and the 1:1 mixture. The theoretical isotherms were calculated from the Frumkin equation (11) for Triton X-100 and $C_{12}EO_5$. The results of calculations from Eq. 62, with a mean molar area of the mixture estimated from Eq. (63), are in a very good agreement with the experimental data.

6.2 Adsorption kinetics of mixed solutions

The general description of the adsorption kinetics of a surfactant mixture can be made in an analogous way as for a single surfactant solution. Instead of Eq. (24) a set of transport equations has to be used, one for each of the r different surfactants. The initial and boundary conditions are defined for each component, in analogy to Eqs. (25) through (28) and finally a system of r integral equations results, either in the form

$$\Gamma_i(t) = 2\sqrt{\frac{D_i}{\pi}}\left(c_{oi}\sqrt{t} - \int_0^{\sqrt{t}} c_i(0, t - \tau)d\sqrt{\tau}\right), \; i = 1,...,r \tag{64}$$

or

$$c_i(0, t) = c_{oi} - \frac{2}{\sqrt{D_i \pi}}\int_0^{\sqrt{t}} \frac{d\Gamma_i(t - \tau)}{dt}d\sqrt{\tau}, \; i = 1,...,r. \tag{65}$$

The set of r integral equations is interlinked via a multi-component adsorption isotherm.

As it was shown above there are simple thermodynamic models that allow a quantitative description of the adsorption behaviour of surfactant mixtures at equilibrium. Using such isotherms a correct description of the adsorption kinetics is also possible, however, additional numerical procedures are required.

The adsorption kinetics of surfactant mixtures is a field rarely investigated and systematic studies do not exist so far (Malqueen et al. 2001, 2001a, Malqueen and Blankschtein 2002). For the thermodynamic model given above Miller et al. (2003) have shown recently some simulations and comparison of experimental data with the theory. Calculated dynamic surface tensions for a solution containing two surfactants are given in Figs. 38 to 40. The parameters for the simulation were selected such that they correspond to the two nonionic surfactants decyl and tetradecyl dimethyl phosiphine oxide. The isotherms of these surfactants were discussed in detail elsewhere (Fainerman et al. 2001). At first, the dynamic surface tensions of the two compounds are shown at the studied concentrations in Fig. 38, each in absence of the second surfactant. Depending on the concentration, the decrease in surface tension starts at respective times. These times of a significant surface tension change can be found in the kinetics of adsorption of the mixture.

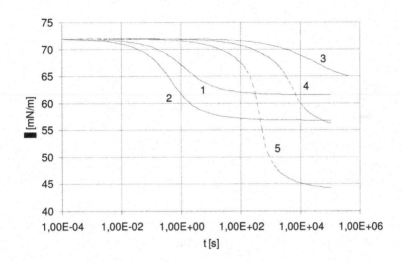

Figure 38. Calculated surface tension γ as a function of time for single solutions of $C_{10}DMPO$ and $C_{14}DMPO$: $c_{10} = 10^{-7}$ (1), $2\ 10^{-7}$ (2), $c_{14} = 10^{-9}$ (3), $3\ 10^{-9}$ (4), 10^{-8} mol/cm³ (5), $D_{10} = D_{14} = 3\cdot10^{-6}$ cm²/s, $\omega_{10} = \omega_{14} = 2.52\cdot10^{9}$ cm²/mol, $b_{10} = 21.9$ m³/mol, $b_{14} = 763$ m³/mol, $a_{10} = -0.22$, $a_{14} = 0.8$

The dependencies $\gamma(t)$ for some mixtures of the two components are displayed in Fig. 39. Until an adsorption time of about 10 s the values of $\gamma(t)$ are almost identical to those calculated for the pure substance $C_{10}DMPO$. If we look at the change in adsorption of the two components, shown in Fig. 40, we see that at about this time the molecules of the $C_{14}DMPO$ really start to adsorb and compete with the adsorbed $C_{10}DMPO$ molecules. Due to this competition, the adsorbed amount of $C_{10}DMPO$ goes through a maximum and then decreases and levels off at the respective equilibrium value. The larger the $C_{14}DMPO$ concentration the earlier appears the maximum in the adsorption of the $C_{10}DMPO$. These model calculations should allow us to understand the adsorption behaviour of mixed $C_{10}DMPO$ and $C_{14}DMPO$ solutions as measured by dynamic surface tensions.

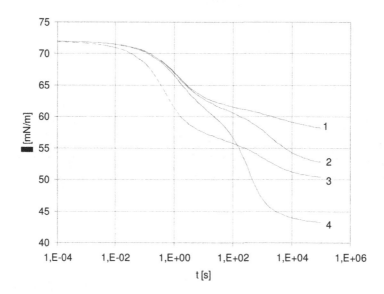

Figure 39. Calculated surface tension γ as a function of time using Eq. (64) to simulate the simultaneous adsorption of mixtures containing $C_{10}DMPO$ and $C_{14}DMPO$: $c_{10} = 10^{-7}$ (1, 2, 3), $2\cdot10^{-7}$ mol/cm³ (4), $c_{14} = 10^{-9}$ (1, 4), $3\cdot10^{-9}$ (2), 10^{-8} mol/cm³ (3), all other parameters as in Fig. 38

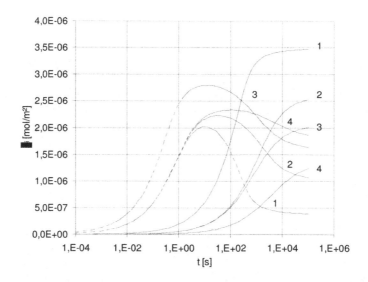

Figure 40. Calculated adsorption Γ as a function of time using Eq. (64) to simulate the simultaneous adsorption of mixtures containing $C_{10}DMPO$ and $C_{14}DMPO$, all other parameters as in Figs. 38 and 39

Note, the course of $\gamma(t)$ is very typical for commercial surfactants which are usually mixtures of homologues or even various different compounds. A very famous example is the most frequently studied surfactant sodium dodecyl sulphate SDS, which is supplied as chemical by the homologous alcohol dodecanol (Vollhardt and Emrich 2000).

6.3 Dynamic surface tensions of a mixed surfactant system

The graph in Fig. 41 shows the dynamic surface tensions of a mixtured solution of $C_{10}DMPO$ and $C_{14}DMPO$ measured with the maximum bubble pressure method BPA1 (O) and profile analysis tensiometer PAT1 ($\square$). The theoretical curves shown were calculated due to the adsorption kinetics model for surfactant mixtures discussed above (Miller et al. 2003).

First of all, we see that the data of the two experimental methods complement each other adequately. The dotted line refers to the diffusion model with a diffusion coefficient of D= 3 10^{-6} cm²/s, which corresponds to the physically reasonable value for this surfactant. One can see that this line does not fit the experimental data very well, however, the solid line does. This solid line was calculated for D = 1 10^{-6} cm²/s, a value slightly smaller than expected. The reason could be that we used here a mixed adsorption isotherm based on a Frumkin model, while it was shown that alkyl dimethyl phosphine oxides are better described by the reorientation model. This could explain the lower value of the diffusion coefficients.

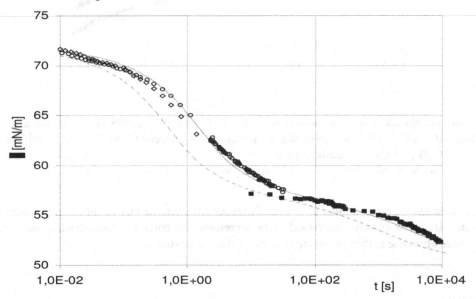

Figure 41. $C_{10}DMPO$ and $C_{14}DMPO$ mixture $2\,10^{-7}$ mol/cm³ | $3\,10^{-9}$ mol/cm³, dotted line - D = 3 10^{-6} cm²/s, solid line – D = 1 10^{-6} cm²/s, calculated using the surfactant parameters given in Fig. 38, open symbols – BPA1, closed symbols – PAT1

In Fig. 42, the dynamic surface tensions for three other mixtures of the two surfactants are presented, together with theoretical curves calculated for different diffusion coefficients, all in the range between $D = 1 \ 10^{-6}$ cm²/s and $D = 3 \ 10^{-6}$ cm²/s. Thus, the diffusion coefficient, taken constant for the two compounds, seems to be in the right physical range and we can conclude that the experiments are well described by the theoretical model. A better description would obviously be possible if the mixed model were based on the reorientation model for both components. For this the kinetic model can be formulated easily, however, a numerical solution does not exist yet.

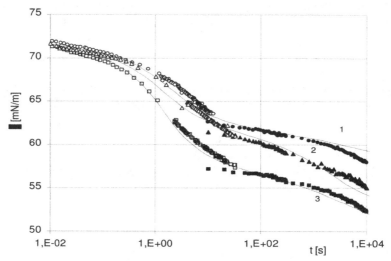

Figure 42. Dynamic surface tensions of a mixture of $C_{10}DMPO$ and $C_{14}DMPO$; concentration ratio $c_{10} | c_{14} = 10^{-7}$ mol/cm³ $| \ 10^{-9}$ mol/cm³ $(\bigcirc \bullet)$, 10^{-7} mol/cm³ $| \ 3 \cdot 10^{-9}$ mol/cm³ $(\triangle \blacktriangle)$, $2 \cdot 10^{-7}$ mol/cm³ $| \ 3 \ 10^{-9}$ mol/cm³ $(\square \blacksquare)$; solid lines - calculated for $D = 1 \cdot 10^{-6}$ (1), $3 \cdot 10^{-6}$ (2) $2 \cdot 10^{-6}$ (3) cm²/ using the surfactant parameters given in Figs. 38 and 39, open symbols – BPA1, closed symbols – PAT1

In conclusion we can state that the thermodynamics and even the kinetics of simple surfactant mixtures are quantitatively understood. This represents the basis for understanding practical surfactant, which due to their purity are typically surfactant mixtures.

References

Aksenenko E.V., A.V. Makievski, R. Miller and V.B. Fainerman, Colloids Surfaces A, 143 (1998a) 311

Aksenenko E.V., V.B. Fainerman and R. Miller, J. Phys. Chem., 102 (1998) 6025

Aniansson E.A.G., S.N. Wall, M. Almgren, H. Hoffmann, I. Kielmann, W. Ulbricht, R. Zana, J. Lang and C. Tondre, J. Phys. Chem., 80 (1976) 905

Aratono M., S. Uryu, Y. Hayami, K. Motomura and R. Matuura, J. Colloid Interface Sci., 98 (1984) 33

Ariel G., H. Diamant and D. Andelman, Langmuir, 15 (1999) 3574

Baret J.F., J. Colloid Interface Sci., 30 (1969) 1

Bashforth F. and J.C. Adams, An Attempt to Test the Theory of Capillary Action, Cambridge University Press and Deighton Bell & Co., Cambridge, 1883

Bendure R.L., J. Colloid Interface Sci., 35 (1971) 238

Bogaert P. van den and P. Joos, J. Phys. Chem., 83 (1979) 2244

Borwankar R.P. and D.T. Wasan, Chem. Eng. Sci., 43 (1988) 1323

Borwankar R.P., L.A. Lobo, D.T. Wasan, Colloids Surfaces A, 69 (1992) 135

Boyd E., J. Phys. Chem., 62 (1958) 536

Brown R.C., Philos. Mag., 13 (1932) 578.

Butler J.A., V. Proc. Roy. Soc. Ser. A, 138 (1932) 348

Chang H.C., C.T. Hsu and S.-Y. Lin, Langmuir, 14 (1998) 2476

Cheng P., D. Li, L. Boruvka, Y. Rotenberg and A.W. Neumann, Colloids Surfaces, 43 (1990) 151

Cheng P., D.Y. Kwok, R.M. Prokop, O.I. del Rio, S.S. Susnar and A.W. Neumann, in "Drops and Bubbles in Interfacial Science", in "Studies in Interface Science", Vol. 6, D. Möbius and R. Miller (Eds.), Elsevier, Amsterdam, 1998, p. 61-138

Daniel R. and Berg J., J. Colloid Interface Sci., 237 (2001) 294

Datwani S.S. and J.K. Stebe, J. Colloid Interface Sci., 219 (1999) 282

Defay R. and I. Prigogin, Surface Tension and Adsorption, Longmans-Green, London, 1966

Doss K.S.G., Koll. Z., 86 (1939) 205

Dukhin S.S., R. Miller and G. Kretzschmar, Colloid Polymer Sci., 263 (1985) 420.

Dushkin C.D., Colloids Surfaces A, 143 (1998) 283

Eastoe J., J.S. Dalton and R.K. Heenan, Langmuir, 14 (1998) 5719.

Fainerman V.B. and R. Miller, Adv. Colloid Interface Sci., (2003), in press

Fainerman V.B. and R. Miller, Colloids Surfaces A, 97(1995)65

Fainerman V.B. and R. Miller, J. Colloid Interface Sci., 175 (1995a) 118

Fainerman V.B. and R. Miller, J. Phys. Chem. B, 105 (2001) 11432

Fainerman V.B. and R. Miller, Langmuir, 12(1996)6011

Fainerman V.B. and R. Miller, The maximum bubble pressure technique, monograph in "Drops and Bubbles in Interfacial Science", in "Studies of Interface Science", D. Möbius and R. Miller (Eds.), Vol. 6, Elsevier, Amsterdam, 1998, p. 279-326.

Fainerman V.B., A.V. Makievski and R. Miller, Colloids Surfaces A, 87 (1994) 61

Fainerman V.B., A.V. Makievski and R. Miller, Labor Praxis, (2003a), in press

Fainerman V.B., D. Möbius and R. Miller (Eds.), Surfactants – Chemistry, Interfacial Properties and Application, in "Studies in Interface Science", Vol. 13, Elsevier, 2001

Fainerman V.B., E.H. Lucassen-Reynders and R. Miller, Colloids Surfaces A, 143 (1998) 141

Fainerman V.B., E.V. Aksenenko and R. Miller, J. Phys. Chem., 104 (2000) 5744

Fainerman V.B., R. Miller and A.V. Makievski, Rev. Sci. Instr., (2003c), in press

Fainerman V.B., R. Miller and E.V. Aksenenko, Adv. Colloid Interface Sci., 96 (2002a) 339

Fainerman V.B., R. Miller and E.V. Aksenenko, Langmuir, 16 (2000a) 4196

Fainerman V.B., R. Miller and H. Möhwald, J. Phys. Chem., 106 (2002) 809

Fainerman V.B., R. Miller and R. Wüstneck, J. Phys. Chem., 101 (1997) 6479

Fainerman V.B., R. Miller, E.V. Aksenenko, A.V. Makievski, J. Krägel, G. Loglio and
    L. Liggieri, Adv. Colloid Interface Sci., 86 (2000a) 83

Fainerman V.B., R. Miller, R. Wüstneck and A.V. Makievski, J. Phys. Chem., 100 (1996) 7669

Fainerman V.B., S.A. Zholob, E.H. Lucassen-Reynders and R. Miller, J. Colloid Interface Sci.,
    261 (2003) 180.

Fainerman V.B., S.V. Lylyk, A.V. Makievski and R. Miller, J Colloid Interface Sci., (2003b), in
    press

Ferrari M., L. Liggieri, F. Ravera, C. Amodio and R. Miller, J. Colloid Interface Sci.,
    186 (1997) 40

Ferrari M., L. Liggieri, F. Ravera, C. Amodio and R. Miller, J. Colloid Interface Sci.,
    186 (1997a) 46

Ferrari M., L. Liggieri and F. Ravera, J. Phys. Chem. B, 102 (1998) 10521

Garrett P.R. and D.R. Ward, J. Colloid Interface Sci., 132 (1989) 475

Hansen R.S. and T. Wallace, J. Phys. Chem., 63 (1959) 1085

Hansen R.S., J. Colloid Sci., 16 (1961) 549

Hansen R.S., J. Phys. Chem., 64 (1960) 637

Hartland S. and R.W. Hartley, Axisymmetric Fluid-Liquid Interfaces, Elsevier Amsterdam, 1976

Joos P., Bull. Soc. Chim. Belg., 76 (1967) 591

Joos P., Dynamic Surface Phenomena, VSP, Utrecht, 1999

Kazakov V.N., O.V. Sinyachenko, V.B. Fainerman, U. Pison and R. Miller, Dynamic Surface
    Tension of Biological Liquids in Medicine, in "Studies in Interface Science", Vol. 8, D. Möbius
    and R. Miller (Editors), Elsevier, Amsterdam, 2000.

Kovalchuk V.I. and S.S. Dukhin, Colloids & Surfaces A, 192 (2001) 131.

Kovalchuk V.I., G. Loglio, V.B. Fainerman and R. Miller, J. Colloid Interface Sci., (2003), in
    press

Kretzschmar G. and K. Lunkenheimer, Ber. Bunsenges. Phys. Chem., 74(1970)1064

Langmuir I., J. Amer. Chem. Soc., 39 (1917) 1848

Lee Y.-C., H.-S. Liu, S.-Y. Lin, Colloids Surfaces A, 212 (2003) 123

Liggieri L., F. Ravera, M. Ferrari, A. Passerone and R. Miller, J. Colloid Interface Sci.,
    186 (1997) 46

Liggieri L., M. Ferrari, A. Massa and F. Ravera, Colloids Surfaces A, 156 (1999) 455

Lin S.-Y., T.-L. Lu and W.-B. Hwang, Langmuir, 10 (1994) 3442

Loglio G., P. Pandolfini, R. Miller, A.V. Makievski, F. Ravera, M. Ferrari and L. Liggieri, in
    "Novel methods to study interfacial layers", Studies in Interface Science, Vol. 11, D. Möbius
    and R. Miller (Eds.), Elsevier, Amsterdam, 2001

Loglio G., R. Miller, A.M. Stortini, U. Tesei, N. Degli Innocenti and R. Cini, Colloids Surfaces A,
    90 (1994) 251

Lucassen J. and M. van den Tempel, Chem. Eng. Sci. 27 (1972) 1283

Lucassen-Reynders E.H., J. Colloid Interface Sci., 41 (1972) 156

Lunkenheimer K. and R. Hirte, J. Phys. Chem., 96 (1992) 8683

Lunkenheimer K. and R. Miller, Material Science Forum, 25 (1988) 351

Lunkenheimer K. and R.Miller, J. Colloid Interface Sci., 120 (1987) 176

Lyklema J., Fundamentals of Interface and Colloid Science, Vol. I, Academic Press, London, 1993

MacLeod C.A. and C.J. Radke, J. Colloid Interface Sci., 166 (1994) 73

Makievski A.V., V.B. Fainerman, R. Miller, M. Bree, L. Liggieri and F. Ravera, Colloids Surfaces A, 122 (1997) 269

Malqueen M. and D. Blankschtein, Langmuir, 15 (1999) 8832

Malqueen M. and D. Blankschtein, Langmuir, 16 (2000)7640

Malqueen M. and D. Blankschtein, Langmuir, 18 (2002) 365

Malqueen M., S.S. Datwani, K.J. Stebe and D. Blankschtein, Langmuir, 17 (2001) 7494

Malqueen M., K.J. Stebe and D. Blankschtein, Langmuir, 17 (2001a) 5196

Malysa K., R. Miller and K. Lunkenheimer, Colloids Surfaces, 53 (1991) 47

Miller R. and G. Kretzschmar, Colloid Polymer Sci., 258 (1980) 85

Miller R., A.V. Makievski, C. Frese, J. Krägel, E.V. Aksenenko and V.B. Fainerman, submitted to Tenside Surfactants Detergents, (2003)

Miller R., Colloids Surfaces, 46(1990)75

Miller R., E.V. Aksenenko, L. Liggieri, F. Ravera, M. Ferrari and V.B. Fainerman, Langmuir, 15 (1999) 1328

Miller R., R. Sedev, K.-H. Schano, Ch. Ng and A.W. Neumann, Colloids Surfaces A, 69 (1993) 209

Miller R., V.B. Fainerman, E.V. Aksenenko, A.V. Makievski, J. Krägel, L. Liggieri, F. Ravera, Wüstneck R. and G. Loglio, in "Emulsions, Foams and Thin Films", P. Kumar and K.L. Mittal (Eds.), Marcel Dekker, 2000, p. 313

Miller R., V.B. Fainerman, K.-H. Schano, W. Heyer, A. Hofmann and R. Hartmann, Labor Praxis, N8(1994)

Milner S.R., Phil. Mag., 13 (1907) 96

Mishchuk N.A., S.S. Dukhin, V.B. Fainerman, V.I. Kovalchuk and R. Miller, Colloids & Surfaces A, 192 (2001) 157.

Mysels K.J., Langmuir, 2 (1986) 428

Mysels K.J., Langmuir, 5 (1989) 442

Neumann A.W. and J.K. Spelt, Applied Surface Thermodynamics, Marcel Dekker, New York, 1996

Patent DE 197 55 291 C1, 1997; Patent DE 199 33 631 A1, 1999.

Penfold J.. E. Staples, I. Tucker, R.K. Thomas, R. Woodling, and C.C. Dong, J. Colloid Interface Sci., 262 (2003) 235

Pratt G. and C. Thoraval, Proceedings of 2nd World Congress on Emulsion, Bordeaux, 1997, 2/2/125/00-2/2/125/05

Ravera F., L. Liggieri and R. Miller, Colloids Surfaces A, 175 (2000) 51

Ravera F., L. Liggieri, A. Passerone and A. Steinchen, J. Colloid Interface Sci., 163 (1994) 309.

Ravera F., M. Ferrari, L. Liggieri, R. Miller and A. Passerone, Langmuir, 13 (1997) 4817

Rotenberg Y., L. Boruvka and A.W. Neumann, J. Colloid Interface Sci. 93 (1983) 169

Rubin E. and C.J. Radke, Chemical Eng. Sci, 35 (1980) 1129

Rusanov A.I. and V.B. Fainerman, Dokl. Akad. Nauk SSSR, 308(1989)651

Rusanov A.I., *Fazovye Ravnovesija i Poverchnostnye Javlenija*, Khimija, Leningrad, 1967

Siddiqui F.A. and E.I. Franses, Langmuir, 12 (1996) 354

Siddiqui F.A. and E.I. Franses, AIChE J., 43 (1997) 1569

Simon M., Ann. Chim. Phys. 32 (1851) 5.

Sugden S., J. Chem. Soc., 121 (1922) 858.

Sutherland K.L., Austr. J. Sci. Res., A5 (1952) 683

Szyszkowski B. von, Z. Phys. Chem. (Leipzig), 64 (1908) 385

Ter Minassian-Saraga L., J. Colloid Sci., 11 (1956) 398

Tian Y., G. Holt and R.E. Apfel, J Colloid Interface Sci., 187 (1997) 1

Ueno M. , Y. Takasawa, H. Miyashige, Y. Tabata and K. Meguro, Colloid Polym. Sci., 259 (1981) 761

Vollhardt D. and G. Emrich, Colloids Surfaces A, 161 (2000) 173

Wantke K.-D. and H. Fruhner, in "Studies in Interface Science", Vol.6, D. Möbius and R. Miller (Eds.), Elsevier Science, Amsterdam, 1998, pp. 327

Ward A.F.H. and L. Tordai, J. Phys. Chem., 14 (1946) 453

Kinetics of spreading of surfactant solutions

V. Starov[*]

Department of Chemical Engineering, Loughborough University, Loughborough, LE11 3TU, UK,
V.M.Starov@lboro.ac.uk

Abstract. Two different mechanism of influence of surfactants on hydrodynamics of spreading are considered: (i) the presence of an inhomogeneous distribution of surfactants on liquid-air interfaces, which results in surface tension gradients, which in their turn cause tangential stresses and Marangoni flow, and (ii) a slow spreading of surfactant solutions over hydrophobic substrates, which is caused by adsorption of surfactant molecules on a bare hydrophobic interface in front of the moving three phase contact line. In section 2 we present results of the theoretical and experimental study of the spreading of an insoluble surfactant over a thin liquid layer. Initial concentrations of surfactant above and below critical micelle concentration (CMC) is considered. If the concentration is above the CMC two distinct stages of spreading are found (a) the fast first stage, which is connected with the micelles dissolution; (b) the second slower stage, when the surfactant concentration becomes below CMC over the whole liquid surface. During the second stage, the formation of a dry spot in the centre of the film is observed. A similarity solution of the corresponding equations for spreading results in the good agreement with the experimental observations. In section 3 the spreading of aqueous surfactant solutions over hydrophobic surfaces is considered from both theoretical and experimental points of view. Aqueous droplets do not wet a virgin solid hydrophobic substrate and do not spread, however, surfactant solutions spread. It is shown that the transfer of surfactant molecules from the aqueous droplet onto the hydrophobic surface changes the wetting characteristics in front of the droplet on the moving three phase contact line. The adsorption of surfactant molecules results in an increase of the solid-vapour interfacial tension and hydrophilisation of the initially hydrophobic solid substrate in front of the spreading droplet. This process causes aqueous droplets to spread over time. The time evolution of the spreading of aqueous droplets is predicted and compared with experimental observations. The assumption that surfactant transfer from the droplet surface onto the solid hydrophobic substrate controls the rate of spreading is confirmed by our experimental observations.

1 Introduction

In section 2 we consider a spreading of insoluble surfactants over thin aqueous layer. When a drop of a surfactant solution is deposited on a clean liquid-air interface, then tangential stresses develop on the liquid surface. They are caused by the non-uniform distribution of the surfactant concentration, Γ, over the liquid surface covered by the surfactant, hence, leading to the creation of a surface stresses and a flow (Marangoni effect) (Levich, 1962):

$$\mu \frac{\partial u(r,h)}{\partial z} = \frac{d\gamma}{d\Gamma} \frac{\partial \Gamma}{\partial r} \tag{1}$$

[*] The author acknowledges the contributions made by the co-workers and the financial support by the Royal Society, UK.

where μ, u are, respectively, the liquid dynamic shear viscosity and tangential velocity on the liquid surface located at height h; (r, z) are radial and vertical coordinates, respectively; $\gamma(\Gamma)$ is the liquid-air interfacial tension whose linear dependency on surfactant surface concentration we assume below. The surface tension gradient-driven flow induced by the Marangoni effect moves surfactant along the surface and a dramatic spreading process takes place. The liquid-air interface in the course of motion deviates from an initially flat position to accommodate the normal stress also occurring.

We restrict consideration here to insoluble surfactants. Note that though a surfactant may be soluble there are cases such that insolubility conditions can be used during a certain short period of the spreading process. Let us consider two characteristic time scales associated with the surfactant transfer: (i) τ_{d*} accounts for the transfer from the liquid-air interface to the bulk, and (ii) τ_{a*} from the bulk back to the interface. In both cases these characteristic time scales depend on an energy difference between corresponding states. For example, if we take water and a surfactant made of a hydrophilic head and a hydrophobic tail, then E_{hb}, E_{ta}, E_{tl} are the energies (in kT units) of head-water, tail-air, and tail-water interactions. Using these notations the energy of a molecule in an adsorbed state at the interface is $E_{ad} = E_{ta} + E_{hl}$, while for the same molecule in the bulk $E_b = E_{tl} + E_{hl}$. Then $\tau_{d*} \sim exp(E_b - E_{ad}) = exp\{E_{tl} - E_{ta}\}$ and $\tau_{a*} \sim exp(E_{ad} - E_b) = exp\{-(E_{tl} - E_{ta})\}$. Generally, the tail-water interaction energy is considerably higher than that of the tail-air interaction ($E_{tl} \approx nE_{tl}^1$, where E_{tl}^1 is an interaction energy per hydrophobic unit and n is a number of those units in each tail).

Consequently, $\dfrac{\tau_{d*}}{\tau_{a*}} \sim exp\{2(E_{tl} - E_{ta})\} >> 1$ and transfer from the interface to bulk is a much slower process than the reverse one. If the duration of a spreading experiment is shorter than τ_{d*}, then during that experiment the surfactant can be considered as insoluble. Otherwise, if $t > \tau_{d*}$ the solubility of the surfactant in the liquid must be taken into consideration. In the latter case surfactant transfer to the bulk liquid tends to make concentration uniform both in the bulk and at the interface, and the result is a substantial decrease of the influence of surfactants.

Usually surface diffusion is neglected as compared to convective transfer. Indeed, from Eq. (1), we have as a characteristic scale of surface velocity: $u_* \approx \dfrac{\gamma_* H_*}{\mu L_*}$, where γ_*, H_*, L_* are characteristic scales of interfacial tension, initial film thickness, and length in a tangential direction, respectively. The diffusion process over the liquid surface scales like: $D_s \dfrac{\Gamma_*}{L_*^2}$, where D_s, Γ_* are a scale of surface diffusion and a characteristic scale of surfactant concentration on the surface, respectively. The ratio of diffusion to convective flux can be estimated as $1/Pe = D_s \mu / \gamma_* H_* \sim 10^{-8} << 1$, for $D_s \sim 10^{-5} cm^2/s$ and

$\gamma_* \approx 10^2 \, dyn/cm$; Pe is the Peclet number. This estimation shows that surface diffusion can be neglected everywhere except for a negligibly small diffusion layer. We disregard it below.

It has been shown in ref. (Starov et al., 1997) that if the liquid layer is thin enough then flow induced by surfactant causes large deformation of the interface and moreover the leading edge of the spreading monolayer forms a shock-like front.

Experiments have clearly shown the formation of a dry spot at the central area of the layer (Gaver and Grotberg 1990; Ahmad and Hansen 1972; Fraaije and Cazabat 1989). This is also one of our findings and it is a part of the results reported here. An attempt to explain the dry spot formation due to the action of surface forces appeared in ref. (Jensen and Grotberg 1992), where the disjoining pressure is taken a function of the film thickness as for non-wettable surfaces. However this is not our case for we have complete wetting. We provide the appropriate explanation also accounting for the expected limited precision of experimental measurements, and for evaporation.

We consider two different cases: (i) when concentration in a droplet of surfactant solution, which is placed in the centre of a liquid film is above critical micelle concentration (CMC), and (ii) when such concentration is below CMC. In the first case the spreading process involves two stages: (i) the faster stage when the surfactant concentration is determined by the dissolution of micelles. This stage yields the maximum attainable surfactant concentration in the film centre, which is independent of time. The duration of that stage is determined by the initial amount of micelles in the drop; (ii) the second slower stage takes place when the surfactant concentration changes on the film centre but the total mass of surfactant remains constant. In both cases a similarity solution provides a power law predicting the position of the moving front $r_f(t)$ as time proceeds (Fig. 1).

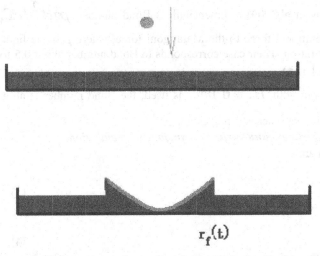

$$r_f(t)$$

Figure 1. Upper part: a drop of surfactant solution is depositing onto the thin aqueous layer. Lower part: a liquid profile during the spreading. Surfactants on the liquid surface are marked in red. $r_f(t)$ is the position of the moving front.

Surfactant adsorption on solid-liquid and liquid-vapour interfaces changes the corresponding interfacial tensions. Liquid motion caused by surface tension gradients on liquid-vapour interfaces (Marangoni effect) is investigated in section 2. The phenomena produced by the presence of surfactant molecules on a solid - vapour interface have been studied less. In section 3 the imbibition of surfactant solutions into thin quartz capillaries was investigated. Spreading of surfactant solutions on both hydrophobic and hydrophilic surfaces (Jensen and Grotberg 1992, Starov t al. 1993, Cachile et al. 1999, Stoebe et al. 1996, 1997, 1997) revealed various intriguing phenomena. In section we address the problem of aqueous surfactant solutions spreading over hydrophobic surfaces from both the theoretical and experimental points of view.

2 The spreading of insoluble surfactants over thin aqueous layers

2.1 Theory and relation to experiment

The motion of a thin liquid layer with initial thickness H_* is considered under the action of an insoluble surfactant on its open surface. For simplicity and in accordance with known data (Starov et al.2000), we assume that surface tension varies linearly with surface concentration of surfactant,

$$\gamma(\Gamma) = \gamma_* - \alpha\Gamma, \quad at \quad 0 < \Gamma < \Gamma_m, \tag{2}$$

where γ_* is the interfacial tension of the pure water-air interface, and Γ_m corresponds to the maximum attainable surface concentration (in equilibrium with a micellar solution).

For the convenience, and universality of our results we use below dimensionless parameters and variables. Expecting no confusion in the reader we use the same symbols as for dimensional quantities. The subscript $*$ is always used to mark initial or characteristic values.

We further assume that $\varepsilon = H_*/L_* \ll 1$, hence neglect of the non-linear part of the

interface curvature. In ref. (Starov et al.2000) a dimensionless Bond number $\rho g H_*^2/\alpha\Gamma_*$

accounts for the ratio of the gravitational force to the Marangoni force, where ρ is the liquid density, and g is the gravity acceleration. Their case corresponds to Bond number about 0.5 for they use a liquid film with $H_* \approx 1\,mm$.

In our experiments a water film with $H_* \approx 0.1mm$ is used, the Bond number is about

$5*10^{-3} \ll 1$ and, hence, the gravity action can be neglected.

Derivation of equations for film thickness and surfactant surface concentration.
For $\varepsilon \ll 1$ the equations of motion are

$$\frac{dp}{dr} = \mu\frac{\partial^2 u}{\partial z^2} \tag{3}$$

$$p = p(r) \tag{4}$$

$$\frac{1}{r}\frac{\partial}{\partial r}(ru) + \frac{\partial v}{\partial r} = 0 \tag{5}$$

where $p(r)$, $v(r,z)$, r, and z are pressure, axial velocity, radial and axial coordinate, respectively. The following boundary conditions must be satisfied:

non-slop condition:

$$u(r,0)=v(r,0)=0 \qquad (6)$$

equality of tangential stresses on the free surface:

$$\mu\frac{\partial u(r,h)}{\partial z}=\frac{\partial \gamma}{\partial r}=\frac{d\gamma}{d\Gamma}\frac{\partial \Gamma}{\partial r}=-\alpha\frac{\partial \Gamma}{\partial r} \qquad (7)$$

and equality of normal stresses:

$$p=p_g-\frac{\gamma}{r}\frac{\partial}{\partial r}\left(r\frac{\partial h}{\partial r}\right) \qquad (8)$$

where p_g is the pressure in the ambient air. Solution of Eqs. (3, 4) with boundary conditions (6), (7) results in

$$u=-\frac{1}{\mu}\frac{\partial}{\partial r}\left(\frac{\gamma}{r}\frac{\partial}{\partial r}\left(r\frac{\partial h}{\partial r}\right)\right)\left(\frac{z^2}{2}-zh\right)-\frac{\alpha}{\mu}\frac{\partial \Gamma}{\partial r}z \qquad (9)$$

Below we make use of a condition at the free liquid-air interface

$$\frac{\partial h}{\partial t}+u(r,h)\frac{\partial h}{\partial r}=v(r,h) \qquad (10)$$

After integration of Eq. (5) over z, from 0 to h, and using the result of integration into Eq. (10) we get the following equation, which describe the evolution of the liquid profile:

$$\frac{\partial h}{\partial t}+\frac{1}{r}\frac{\partial}{\partial r}\int_0^h rudz=0 \qquad (11)$$

In a similar way using the mass balance of surfactant molecules on the liquid interface we arrive to:

$$\frac{\partial \Gamma}{\partial t}+\frac{1}{r}\frac{\partial}{\partial r}(ru(t,h)\Gamma)=0 \qquad (12)$$

where $h(t,r)$ is the film thickness at time t; r is the radial coordinate, and $\Gamma(t,r)$ the surfactant concentration on the surface. Using expression (9) for he velocity profile in Eqs. (11)-(12) and after performing the integration we conclude:

$$\frac{\partial h}{\partial t} = -\frac{1}{r}\frac{\partial}{\partial r}\left\{r\left[\frac{\beta h^3}{3}\frac{\partial}{\partial r}\left(\frac{\gamma}{r}\frac{\partial}{\partial r}\left(r\frac{\partial h}{\partial r}\right)\right) - \frac{h^2}{2}\frac{\partial \Gamma}{\partial r}\right]\right\}$$ (13)

$$\frac{\partial \Gamma}{\partial t} = -\frac{1}{r}\frac{\partial}{\partial r}\left\{r\Gamma\left[\frac{\beta h^2}{2}\frac{\partial}{\partial r}\left(\frac{\gamma}{r}\frac{\partial}{\partial r}\left(r\frac{\partial h}{\partial r}\right)\right) - h\frac{\partial \Gamma}{\partial r}\right]\right\}$$ (14)

that are to be solved subject to the following boundary conditions:

$$\frac{\partial h}{\partial r} = \frac{\partial^3 h}{\partial r^3} = 0, \; at \quad r = 0,$$ (15)

$$h \to 1, \quad at \quad r \to \infty$$ (16)

$$\frac{\partial \Gamma}{\partial r} = 0, \quad at \quad r = 0$$ (17)

$$\Gamma \to 0, \quad at \quad r \to \infty$$ (18)

Let us introduce the following dimensionless variables and values:

$$h \to \frac{h}{H*}, \quad r \to \frac{r}{L*}, \quad \Gamma \to \frac{\Gamma}{\Gamma*}, \quad t \to \frac{t}{t*},$$

$$t* = \frac{\mu L_*^2}{H*\alpha\Gamma*}, \quad \beta = \varepsilon^2\frac{\gamma*}{\alpha\Gamma*} << 1$$

The time scale t_* deserves a comment. Although such time scale is useful for some calculations, however, in the case under consideration it is much shorter than the time scale, which governs the process under consideration. Indeed, the capillary number for the spreading process is very small: $Ca = \frac{\mu U}{\gamma_*} \sim 10^{-2}\,10^{-1}/10^2 = 10^{-5} << 1$. On the other hand

$$Ca = \frac{\mu U_*}{\gamma_*} = \frac{\mu}{\gamma_*}\frac{L_*}{\tau_*}, \text{ hence } \tau_* = \frac{\mu L_*}{\gamma_* Ca},$$ where τ_*, U_* are, respectively, the time scale,

which actually governs the spreading process and its characteristic velocity scale. Hence, we have $\dfrac{t_*}{\tau_*} = \dfrac{\gamma_* Ca}{\alpha \Gamma_* \varepsilon} = \delta \approx 10^{-3} \ll 1$ for $\varepsilon \approx 10^{-2}$, as it is in experiments below. If we now

introduce the dimensionless time $\tau = \dfrac{t}{\tau_*}$, then in Eqs. (13-14) we have $\delta \dfrac{\partial}{\partial \tau} = \dfrac{\partial}{\partial t}$. As we

show below the time evolution of the position of the moving film front is $r_f(t) \sim$

$t^{0.5 \div 0.25} = \left(\dfrac{\tau}{\delta}\right)^{0.5 \div 0.25}$. If we take into account the initial value, ℓ, of $r_f(t)$ this dependence

takes the form $r_f(\tau) \approx \left(\dfrac{\tau}{\delta} + \ell\right)^{0.5 \div 0.25}$, where $\ell \sim 1$ represents the contribution of the initial

condition. According to our choice $\tau \sim 1$, hence, $\dfrac{\tau}{\delta} \gg \ell$, and thus $r_f(t) = \left(\dfrac{\tau}{\delta}\right)^{0.5 \div 0.25} =$

$t^{0.5 \div 0.25}$. This is an interesting result and it justifies why a power law works in spite of the violation of the initial condition.

Multiplying Eq. (13) by r we conclude after integration:

$$\int_0^\infty r(h-1)dr = 0 \qquad\qquad (19)$$

which is the conservation law for the liquid.

In experiments below a droplet of surfactant solution with concentration above CMC was placed in the centre of a water film. Experimental observations showed that two distinct stages of spreading take place: (i) a first faster stage, and (ii) a second slower stage. During the first stage, there is dissociation of micelles, hence, surface concentration in the centre is kept constant during that stage and $\Gamma(t,0) = \Gamma_m$. Choosing $\Gamma_* = \Gamma_m$ as a characteristic scale for surfactant concentration we have in dimensionless form the following boundary condition during the first stage

$$\Gamma(t,0) = 1 \qquad\qquad (20)$$

The first stage lasts until all micelles are dissolved. The duration t_{1*} of that stage is considered below. After the instant t_{1*} a second stage starts. During the second stage the total mass of surfactant remains constant, hence the following boundary conditions apply

$$\dfrac{\partial \Gamma}{\partial r} = 0, \quad at \quad r = 0, \qquad\qquad (21)$$

and

$$\int_0^\infty r\Gamma dr = 1,$$

$$(22)$$

where the characteristic scale for Γ is $\dfrac{Q_*}{2\pi L_*^2}$, with Q_* being the total amount of surfactant on the surface of the droplet.

The first spreading stage. The spreading process in this case is described by Eqs. (13-14) with boundary conditions (15-18, 20). According to section 2.3 the influence of capillary forces can be neglected for $t \gg \beta$, and Eqs. (13-14) become

$$\frac{\partial h}{\partial t} = \frac{1}{r}\frac{\partial}{\partial r}\left\{r\left[\frac{h^2}{2}\frac{\partial \Gamma}{\partial r}\right]\right\}$$

$$(23)$$

$$\frac{\partial \Gamma}{\partial t} = \frac{1}{r}\frac{\partial}{\partial r}\left\{r\Gamma\left[h\frac{\partial \Gamma}{\partial r}\right]\right\}$$

$$(24)$$

Eqs. (23-24) can not satisfy the boundary conditions at $r \to \infty$ and a shock-like spreading front forms (for the derivation of these conditions see section 2.4).

In our case $h_+ = 1$, $\Gamma_+ = 0$, hence, $\dfrac{\partial \Gamma_+}{\partial r} = 0$. Then using conditions (62-63) we get:

$$\dot r f(h_- - 1) = -\frac{1}{2}h_-^2\frac{\partial \Gamma_-}{\partial r}$$

$$(25)$$

$$\Gamma_-(\dot r f + h_-\frac{\partial \Gamma_-}{\partial r}) = 0$$

$$(26)$$

Eq. (26) actually implies two conditions:

$$\Gamma_- = 0$$

$$(27)$$

$$\dot r f = -h_-\frac{\partial \Gamma_-}{\partial r}$$

$$(28)$$

The matching of asymptotic expansions at the moving shock front (see section 2.5) shows that both conditions (27, 28) must be satisfied.

Let us now introduce a new variable $\xi = \dfrac{r}{t^{1/2}}$. Then we have

$$h(t,r) = f(\xi), \quad \Gamma(t,r) = \varphi(\xi) \tag{29}$$

where $0 < r < r_f(t) = v\sqrt{t}$; the constant v is determined below. In condition (19) the upper limit of integration must be replaced now by $r_f(t) = v\sqrt{t}$, hence

$$\int_0^v \xi f(\xi)d\xi = \frac{v^2}{2} \tag{30}$$

which is compatible with definitions (29). Eqs. (23-24) using definitions (29) become:

$$-\xi f'(\xi) = \frac{1}{\xi}\left(\xi f^2(\xi)\varphi'(\xi)\right)' \tag{31}$$

$$-\frac{1}{2}\xi\varphi'(\xi) = \frac{1}{\xi}\left(\xi f(\xi)\varphi(\xi)\varphi'(\xi)\right)' \tag{32}$$

with the corresponding boundary conditions, which follow from (25-28). We have

$$v[f(v)-1] = -f^2(v)\,\varphi'(v),$$

$$\varphi(v)=0,$$

$$\frac{v}{2} = -f(v)\,\varphi'(v).$$

The latter boundary conditions after simple transformations become:

$$\varphi(v) = 0,$$

$$f(v) = 2, \tag{33}$$

$$\varphi'(v) = -\frac{v}{4}.$$

If we again change variables using

$$\eta = \frac{\xi}{v}, \quad f = \Psi(\eta), \quad \varphi = v^2 G(\eta) \tag{34}$$

then the new functions Ψ and G satisfy the same Eqs. (31, 32) with the variable η, where $0 < \eta < 1$, and the following boundary conditions at $\eta = 1$

$$\Psi(1) = 2, \; G(1) = 0 \text{ and } G'(1) = -1/4 \tag{35}$$

Hence, the problem under consideration does not depend on the unknown parameter v.

Calculated dependencies of dimensionless film thickness $\Psi(\eta)$ and surface concentration $G(\eta)$ are presented in Fig. 2, which shows that a substantial depression is formed in the film centre. During the second stage the depression becomes a dry spot right in the middle of the film.

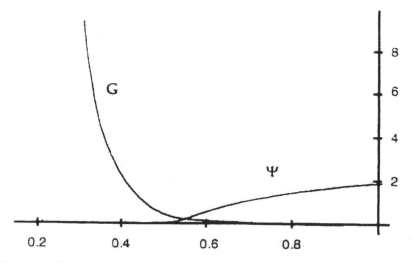

Figure 2. Theoretical predictions for dimensionless profile $\Psi(\eta)$ and surfactant surface concentration $G(\eta)$ during the first stage of spreading.

In order to determine the unknown value v let us consider the total mass of surfactant $Q(t)$ during the first spreading stage. We have

$$Q(t) = 2\pi \int_0^\infty r\Gamma(t,r)dr = 2\pi L_*^2 \Gamma_* t \int_0^v \xi\varphi(\xi)d\xi$$

or

$$\int_0^v \xi\varphi(\xi)d\xi = \frac{Q(t)}{2\pi L_*^2 \Gamma_* t} \tag{36}$$

The left hand side of Eq. (36) does not depend on time t, hence, the same is true for the right hand side. Let us denote by q a constant value to be experimentally determined from the duration of the first stage. Then, Eq. (36) takes the following form:

$$\int_0^v \xi\varphi(\xi)d\xi = q \tag{37}$$

On the other hand, according to definition (34), $v^4 \int_0^1 \eta G(\eta)d\eta = q$. The latter equation should give a possibility to determine v. Unfortunately the integral $\int_0^1 \eta G(\eta)d\eta$ diverges due to the singularity of $G(\eta)$ at $\eta = 0$. Hence, we can only conclude that $v^4 \sim q$, or:

$$v \approx q^{1/4} \tag{38}$$

Let t_{1*} be a dimensional time characterizing the duration of the first spreading stage; t_{1*}/t_* is the corresponding dimensionless time. If we choose $t_* = t_{1*}$ then using the definition of the time scale we get the corresponding value of the tangential length scale

$$L_* = \left(\frac{t_{1*}H_*\alpha\Gamma_*}{\mu}\right)^{1/2}.$$

From Eq. (36) we conclude:

$$q = \frac{Q_*}{2\pi L_*^2 \Gamma_*} \tag{39}$$

where Q_* is the total amount of surfactant which is initially placed on the film surface, which is supposed to be fixed by the experimental conditions, $Q_* = V_* C_*$, where V_* and C_* are the droplet volume and surfactant concentration in the droplet respectively.

Unfortunately, the derived similarity solution does not satisfy boundary condition (20) at the origin as the concentration dependence on radial coordinate diverges in a vicinity of the origin. Thus it is more convenient to redefine a characteristic scale of surfactant surface concentration from Eq. (37) using the condition $q=1$ in Eq. (37) (this choice gives the same characteristic scale during both stages of the spreading process). The v value is still undetermined but we give below a way of its calculation.

At the instant t_{1*} the second stage of the spreading process starts.

The second stage of spreading. During this stage the film profile and the surfactant concentration obey the same system of Eqs. (23, 24) with boundary conditions (25-28, 21, 22). Below we assume that γ is a constant and that constant is included in β.

Let us introduce $\xi = \dfrac{r}{t^{1/4}}$, then the solution of Eqs. (23, 24) is

$$h(t,r) = f(\xi), \quad \Gamma(t,r) = \frac{\varphi(\xi)}{t^{1/2}},$$ where two unknown functions $f(\xi)$ and $\varphi(\xi)$ obey

the following system of equations

$$\xi^2 f'(\xi) = \left\{ \xi \left[\frac{4\beta f^3(\xi)}{3} \left(\frac{1}{\xi} (\xi f'(\xi))' \right)' - 2f^2(\xi)\varphi'(\xi) \right] \right\}' \tag{40}$$

$$(\xi^2 \varphi(\xi))' = \left\{ \xi\varphi(\xi) \left[2\beta f^2(\xi) \left(\frac{1}{\xi} (\xi f'(\xi))' \right)' - 4f(\xi)\varphi'(\xi) \right] \right\}' \tag{41}$$

Eq. (41) can be integrated (integration constant is obviously equal to zero), which gives:

$$\xi\varphi(\xi) \left\{ \xi - \left[2\beta f^2(\xi) \left(\frac{1}{\xi} (\xi f'(\xi))' \right)' - 4f(\xi)\varphi'(\xi) \right] \right\} = 0$$

Thus, either

$$\varphi(\xi) = 0 \tag{42}$$

or

$$\varphi'(\xi) = -\frac{\xi}{4f(\xi)} + \frac{\beta}{2} f(\xi) \left(\frac{1}{\xi} (\xi f'(\xi))' \right)'$$ (43)

In the first case, from Eq. (40) we conclude that

$$\xi^2 f'(\xi) = \frac{4\beta}{3} \left\{ \xi f^3(\xi) \left(\frac{1}{\xi} (\xi f'(\xi))' \right)' \right\}',$$ (44)

which is valid at the periphery of the spreading part of the film.

Eq. (44) describes decaying capillary waves on the film surface. Indeed, if we introduce a new local variable near the moving edge λ (to be defined below), $\varsigma = (\xi - \lambda)/\chi$, with

$$\chi = \left(\frac{4\beta}{3\lambda} \right)^{1/3}$$. Using this variable we get from Eq. (44): $f'''(\varsigma) = \frac{f(\varsigma) - 1}{f^3(\varsigma)}$

The asymptotic behaviour of the latter equation yields:

$$f(\varsigma) \approx 1 + e^{-\frac{\varsigma}{2}} (A_1 \cos \frac{\sqrt{3}\varsigma}{2} + A_2 \sin \frac{\sqrt{3}\varsigma}{2}), \quad at \quad \varsigma \to \infty,$$

which describes decaying capillary waves ahead of the advancing front.

In the opposite case, when Eq. (43) is valid we obtain from Eq. (40) using Eq. (43) that

$$\xi^2 f'(\xi) = \left\{ \xi \left[\frac{\beta f^3(\xi)}{3} \left(\frac{1}{\xi} (\xi f'(\xi))' \right)' + \frac{f(\xi)\xi}{2} \right] \right\}'$$ (45)

The value of λ is determined as a point where $\varphi(\lambda) = 0$, or from Eq. (43) and condition (21):

$$2 = \int_0^\lambda \left(\frac{\xi}{4f(\xi)} - \frac{\beta}{2} f(\xi) \left(\frac{1}{\xi} (\xi f'(\xi))' \right)' \right) d\xi$$ (46)

The solution of Eq. (43) is:

$$\varphi(\xi) = \int\limits_{\xi}^{\lambda} \left[\frac{\xi}{4f(\xi)} - \frac{\beta}{2} f(\xi) \left(\frac{1}{\xi} (\xi f'(\xi))' \right) \right]' d\xi,$$

where the integration constant is zero according to Eq. (46).

It is easy to find a solution of the problem under consideration in the zero approximation by setting $\beta = 0$ in Eqs. (43, 44, 46) (section 2.6). It is also possible to calculate an expression for λ using a zero order solution from section 2.6 and Eqs. (43, 44, 46), which gives $\lambda \approx 2^{5/4}$.

In conclusion of the theoretical part let us summarize the obtained results. The dimensional radius of the moving axisymmetric front during the first stage of spreading is given by the following dependence:

$$r_f = v L_* \left(\frac{t}{t_{1*}} \right)^{1/2} \quad (cm), \quad t \le t_{1*} \tag{47}$$

with v still undetermined. During the second stage:

$$r_f = \lambda L_* \left(\frac{t}{t_{1*}} \right)^{1/4} \quad (cm), \quad \lambda \approx 2^{5/4}, \quad t \ge t_{1*} \tag{48}$$

Note that t_{1*} (sec) is the duration of the first stage of spreading, and

$$L_* = \left(\frac{H * \alpha \Gamma * t_{1*}}{\mu} \right)^{1/2}, \quad \Gamma_* = \frac{Q_*}{2\pi L_*^2}.$$

As the front position must be the same at time t_{1*} according to both Eqs. (47) and (48), then $v = \lambda$, and Eq. (47) becomes

$$r_f = \lambda L_* \left(\frac{t}{t_{1*}} \right)^{1/2} \quad (cm), \quad t \le t_{1*} \tag{49}$$

Our theory predicts that the layer thickness decreases in the centre with a vanishing value at the origin, which is the dry spot. When comparing with experimental data we must take into account both the finite precision in the measurements of the film thickness and the possibility of the evaporation during the duration of the experiment. Both may result in a discrepancy between our theory predictions and the experimental measurements. Indeed, the film thickness can be measured only at values higher than a certain thickness h_*, hence, according to that al thicknesses below h_* constitute the dry spot from the experimental point of view. Using a zer order solution (76) we conclude:

$$2H_*\left(\frac{r_d}{\lambda t^{1/4}}\right)^2 = h_*, \quad or \quad r_d(t) = \lambda\left(\frac{h_*}{2H_*}\right)^{1/2} t^{1/4}, \tag{50}$$

where $r_d(t)$ is the radius of "the dry spot", whose motion according to the latter equatio obeys the same power law as the front of the film.

On the top of that, the evaporation has more pronounced influence at smaller thicknesse $(h<h_*)$ and that influence progressively increases with time. Hence, the evaporation in th liquid film near the centre of the layer enhance the further thinning of the film during botl stages.

2.2 Experimental results

Observations of the spreading of a surfactant on a thin layer of liquid have been performe using aqueous solutions of sodium dodecyl sulfate (SDS) at a concentration $c = 20$ g/l abov CMS (the critical micelle concentration of the SDS is 4 g/l) (Starov et al., 1997). A thin laye of liquid was formed by coating the bottom of a borosilicate glass Petri dish of diameter 20 c using 10 ml of distilled water. The resulting thickness was $H = 0.32 \pm 0.01$ mm. This layer di not dewet during the duration of the experiment. A drop of the surfactant solution, volume 3 μ was placed on the top of the surface of this water layer using a syringe. After touching th water surface, the surfactant spreads on it and this motion monitored using a small amount o talc powder as a marker, and a 25 Hz-video camera to record the time evolution.

The spreading of surfactants forced the water to flow away from the initial location of th drop thus creating a depression where only a thin film of liquid subsisted. The periphery of thi depression, i.e. the liquid front had a sharp increase in thickness. The water layer wa horizontal and the drop was carefully placed, hence, there was no preferred direction, and th edge was circular but some modulation could appear in the experiment after a few seconds Note that the surfactant occupied more surface area of the film than the depressed zone sinc the talc powder is pushed ahead of it. The dependence on time of the radius of the surfactan patch is given in Fig. 3 using log-log plot.

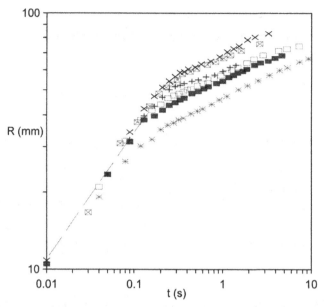

Figure 3. Radius of the spreading front versus time. Points correspond to six different experimental runs.

The two successive stages earlier predicted (section 2.1) are clearly seen in Fig. 3. First, the short period when the surfactant spreads following the power-law: $r_f(t) \sim t^{0.60 \pm 0.15}$. The exponent was not determined with a high precision because the duration of the first stage was too short (about *0.1 s*), which was only three times our time resolution (*0.04 s*). However, the experimental value of he exponent, 0.60±0.15, agrees well with the theoretical prediction, 0.5, according to Eq. (49).

At the end of the first stage the motion of the front abruptly slowed down and the moving front followed a different power law $r_f(t) \sim t^{0.17 \pm 0.02}$. The new exponent is smaller than the theoretically predicted value, 0.25, given by Eq. (48). In Fig. 4, the radius of the shallow region, i.e. the dry spot radius is plotted for the same experimental runs as above.

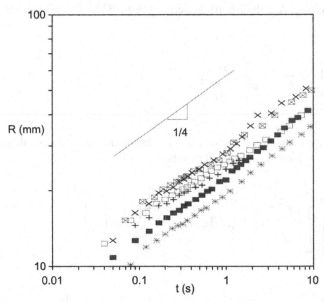

Figure 4. Radius of the dry spot versus time. Points correspond to six different experimental runs.

The observed power law, $r_d(t) \sim t^{0.25 \pm 0.05}$, is in the good agreement with the value predicted by the theory, according to Eq. (50). According to measurements the ratio $\dfrac{r_d(t)}{r_f(t)}$ is about 1/3 although it slowly changes with time during the second stage. Thus, from Eq. (50) we conclude that $h_* \approx 0.07$ cm and it is the lowest thickness which can be detected by experimental method used in (Starov et al., 1997).

In conclusion, the radius of the dry sport moves with the speed predicted by the theory while the front edge of the moving part of the layer proceeds slower than theoretically predicted during the second spreading stage. The discrepancy may be due to one ore both of the following reasons:

(i) gravity action, which is much more pronounced at the front higher edge of the layer than at the lower edge (dry spot). Although the Bond number is very low in our experiments, flow reversal onset cannot be ruled out during the second stage;

(ii) our assumption that the surfactant used is insoluble during the whole duration of the experiment may not be fully correct. Unfortunately we have been unable to estimate the time scale of the desorption, τ_d. If the duration of the experiment is comparable with τ_d then the desorption of surfactant from the liquid surface into the bulk may be significant during the second stage at the front edge.

2.3 Influence of capillary forces during the initial stage of spreading.

Let us estimate the influence of capillary forces during the very short initial stage of spreading where it is significant. Later on it can be neglected everywhere except for thin boundary layers which we consider negligible. A solution of governing Eqs. (13-14) is assumed in the following form:

$$h = f\left(\frac{r}{t^\omega}\right), \quad \Gamma = \varphi\left(\frac{r}{t^\omega}\right)$$

(51)

where f and φ are two new unknown functions; the exponent ω is to be determined. Substitution of expressions (51) into Eqs. (13-14) results in

$$\frac{\omega}{t}\xi f'(\xi) = \frac{1}{\xi}\left\{\xi\left[\frac{\beta f^3(\xi)}{3t^{4\omega}}\left(\frac{\gamma}{\xi}(\xi f'(\xi))'\right)' - \frac{f^2(\xi)}{2t^{2\omega}}\varphi'(\xi)\right]\right\}'$$

(52)

$$\frac{\omega}{t}\xi\varphi'(\xi) = \frac{1}{\xi}\left\{\xi\varphi(\xi)\left[\frac{\beta f^2(\xi)}{2t^{4\omega}}\left(\frac{\gamma}{\xi}(\xi f'(\xi))'\right)' - \frac{f(\xi)}{t^{2\omega}}\varphi'(\xi)\right]\right\}',$$

(53)

where $\xi = r/t^\omega$. There are two ways to choose ω.

(i)   if we require $t = t^{4\omega}$ or $\omega = 1/4$, then Eqs. (51-52) become

$$\frac{1}{4}\xi f'(\xi) = \frac{1}{\xi}\left\{\xi\left[\frac{\beta f^3(\xi)}{3}\left(\frac{\gamma}{\xi}(\xi f'(\xi))'\right)' - t^{1/2}\frac{f^2(\xi)}{2}\varphi'(\xi)\right]\right\}'$$

(54)

$$\frac{1}{4}\xi\varphi(\xi) = \frac{1}{\xi}\left\{\xi\varphi(\xi)\left[\frac{\beta f^2(\xi)}{2}\left(\frac{\gamma}{\xi}(\xi f'(\xi))'\right)' - t^{1/2}f(\xi)\varphi(\xi)\right]\right\}'$$

(55)

Eqs. (54-55) show that the influence of the surfactant grows with time.

(ii)   if we require $t = t^{2\omega}$ or $\omega = 1/2$, then Eqs. (52-53) become

$$\frac{1}{2}\xi f'(\xi) = \frac{1}{\xi}\left\{\xi\left[\frac{\beta f^3(\xi)}{3t}\left(\frac{\gamma}{\xi}(\xi f'(\xi))'\right)' - \frac{f^2(\xi)}{2}\varphi'(\xi)\right]\right\}' \tag{56}$$

$$\frac{1}{2}\xi \varphi'(\xi) = \frac{1}{\xi}\left\{\xi\varphi(\xi)\left[\frac{\beta f^2(\xi)}{2t}\left(\frac{\gamma}{\xi}(\xi f'(\xi))'\right)' - f(\xi)\varphi'(\xi)\right]\right\}' \tag{57}$$

Eqs. (56) and (57) show that the influence of capillary forces decay with time. According to Eq. (51) the spreading law is $r_f(t) \approx t^{1/4}$ during the time period when the capillary force influence is dominant, and $r_f(t) \approx t^{1/2}$ during the time period when the influence of the surfactant gradient is dominant. Here, $r_f(t)$ is the location of the moving boundary of the spreading front. It follows from Eqs. (54-55) that capillary force influence is dominant if $\beta >> t^{1/2}$ or:

$$t << \beta^2 \tag{58}$$

In the same way from (56-57) we find that the capillary force influence is negligible and the influence of the surfactant is dominant if

$$t >> \beta \tag{59}$$

Thus, the capillary force influence is significant during a very short time interval only $t << \beta^2$. As $\beta<<1$, and $t << \beta^2 << \beta$ we can safely consider just the asymptotic behavior when the surfactant influence is dominant and condition (59) is satisfied. Note that by omitting the capillary force action we neglect the highest derivatives in Eqs. (13-14), hence, thin boundary layer arises (in a vicinity of the moving front), where the capillary force action is of the same order magnitude as the surfactant action.

2.4 Derivation of boundary condition at the moving shock front.
Multiplication of Eqs. (23-24) by r and integration over r from r_1 to r_2, where $r_1 < r_f(t) < r_2$, here r_1, r_2 are some constant values, yields

$$\frac{d}{dt}\int_\eta^{r_2} rh(t,r)dr = \frac{r_2 h_2^2}{2}\frac{\partial \tau_2}{\partial r} - \frac{\eta h_1^2}{2}\frac{\partial \tau_1}{\partial r} \tag{60}$$

$$\frac{d}{dt}\int_{r_1}^{r_2} r\Gamma(t,r)dr = r_2\Gamma_2 h_2 \frac{\partial r_2}{\partial r} - r_1\Gamma_1 h_1 \frac{\partial r_1}{\partial r}$$

(61)

where $f_i = f(r_i)$. The left hand side of Eqs. (60) and (61) can be transformed in the following way:

$$\frac{d}{dt}\int_{r_1}^{r_2} rf(t,r)dr = \frac{d}{dt}\left(\int_{r_1}^{r_f(t)} rf(t,r)dr + \int_{r_f(t)}^{r_2} rf(t,r)dr \right) =$$

$$\dot{r}_f r_f f_- + \int_{r_1}^{r_f(t)} r\frac{\partial f(t,r)}{\partial t}dr - \dot{r}_f r_f f_+ + \int_{r_f(t)}^{r_2} r\frac{\partial f(t,r)}{\partial t}dr$$

where $f_\pm = f(t,r_f\pm)$. If now we consider limits r_1 tends to r_f from below (↑) and r_2 tends to r_f from above (↓) when both integrals in the left hand side of the latter equation vanish. Then, from Eqs. (60-61) using the same limits $r_1 \uparrow r_f, r_2 \downarrow r_f$ we conclude:

$$\dot{r}_f(h_- - h_+) = \frac{1}{2}(h_+^2 \frac{\partial \Gamma_+}{\partial r} - h_-^2 \frac{\partial \Gamma_-}{\partial r})$$

(62)

$$\dot{r}_f(\Gamma_- - \Gamma_+) = \Gamma_+ h_+ \frac{\partial \Gamma_+}{\partial r} - \Gamma_- h_- \frac{\partial \Gamma_-}{\partial r}$$

(63)

which are the required boundary conditions at the shock front.

2.5 Matching of asymptotic solutions at the moving shock front.

Let us introduce a new local variable $\varsigma = \frac{\xi - v}{\chi(t)}$, where $\chi(t) \ll 1$ is a new unknown length scale to be determined below. Neglecting the second curvature in Eqs. (13-14) and introducing unknown functions in the following form $\Gamma = \chi(t)\Phi(\xi), h = F(\varsigma)$ we get

$$\chi(t) = \left(\frac{\beta}{t}\right)^{1/3} \tag{64}$$

It follows from the latter equation that $\chi(t) \to 0$, at $t \to \infty$. Unknown functions $\Phi(\varsigma)$ and $F(\varsigma)$ obey the following equations

$$vF' = \left(\frac{2F^3 F'''}{3} - F^2 \Phi'\right)' \tag{65}$$

$$v\Phi' = \left(\Phi F^2 F''' - 2F\Phi\Phi'\right)' \tag{66}$$

After integration of Eqs. (65-66) with boundary conditions $F \to 1$, $\Phi \to 0$, at $\varsigma \to \infty$ we get

$$v(F-1) = \frac{2}{3}F^3 F''' - F^2 \Phi' \tag{67}$$

$$\Phi\left(v - F^2 F''' + 2F\Phi'\right) = 0 \tag{68}$$

Thus, either

$$\Phi = 0 \tag{69}$$

$$F''' = \frac{3v}{2}\frac{F-1}{F^3}$$

or

$$F''' = \frac{3v(F-2)}{F^3} \tag{70}$$

and

$$\Phi' = -\frac{v(3-F)}{F^2} \tag{71}$$

From Eqs. (70-71) we conclude that

$$F \rightarrow 2, \quad \Phi' \rightarrow -\frac{v}{4}, \quad at \quad \varsigma \rightarrow -\infty \tag{73}$$

Eq. (73) shows that the boundary conditions (32) at the shock front are the only possible conditions which can be matched with the inner solution. From the above derivation we conclude that in the boundary layer $\Gamma = \left(\dfrac{\beta}{t}\right)^{1/3} \Phi(\varsigma) \sim \beta^{1/3}$, hence, vanishes from the point of view of the outer solution. Thus, both conditions (26) and (27) must be satisfied at the shock front.

2.6 Solution of the governing equations for the second stage of spreading.
Let us set $\beta=0$ in Eqs. (43, 44), which yields:

$$f'(\xi) = \frac{2f(\xi)}{\xi}, \quad \varphi'(\xi) = -\frac{\xi}{4f(\xi)} \tag{74}$$

with boundary conditions which follow from section 2.4 Eqs. (62, 63)

$$f(\lambda) = 2, \quad \varphi(\lambda) = 0, \quad \varphi'(\lambda) = -\frac{\lambda}{8} \tag{75}$$

Then the solution is

$$f(\xi) = 2\left(\frac{\xi}{\lambda}\right)^2, \quad \varphi(\xi) = -\frac{\lambda^2}{8} \ln \frac{\xi}{\lambda} \tag{76}$$

Substitution of $f(\xi)$ in Eq. (45) gives

$$\lambda = 2^{5/4} \tag{77}$$

Note that the solution (76) satisfies Eqs. (43, 44, 45) at arbitrary β.

3   Spreading of surfactant solutions over hydrophobic substrates
3.1   Theory
Let a small water drop be placed on a hydrophobic surface (Fig 5).

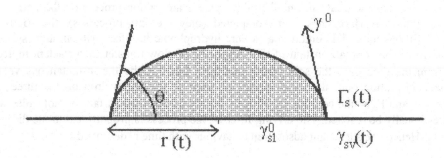

Figure 5. Sketch of the geometry of a drop placed on a solid substrate

If the drop is small enough then the effect of gravity can be ignored. Accordingly, the drop radius r has to be smaller than the capillary length, a, and hence, $r \leq a = \sqrt{\dfrac{\gamma}{\rho g}}$, where ρ, and γ are the liquid density and liquid-vapor interfacial tension, respectively; g is the gravity acceleration.

Let us assume that in the absence of surfactant the drop forms an equilibrium contact angle above $\pi/2$. If the water contains surfactants then three transfer processes take place from the liquid onto all three interfaces: surfactant adsorption at both (i) the inner liquid-solid interface and (ii) the liquid-vapor interface, and (iii) transfer from the drop onto the solid-vapor interface just in front of the drop. Adsorption processes (i) and (ii) result in a decrease of corresponding interfacial tensions, γ_{sv} and γ. The transfer of surfactant molecules onto the solid-vapour interface in front of the drop results in an increase of a local free energy, however, the total free energy of the system decreases. That is, surfactant molecule transfer (iii) goes via a relatively high potential barrier and, hence, goes considerably slower than adsorption processes (i) and (ii). Hence, they are "fast" processes as compared with the third process (iii).

All three surfactant transfer processes are favorable to spreading, as they result in both an increase of the spreading power, $\gamma_{sv} - \gamma - \gamma_{sl}$, and, hence, a decrease of the contact angle (Fig. 5). As it was mentioned above, the transfer of surfactant molecules from the drop onto the solid-vapor interface in front of the drop results in an increase of local surface tension, γ_{sv}. Hence, it is the slowest process that will be the rate determining step. Let us define the initial contact angle by

$$\cos\theta^0 = \frac{\gamma_{sv}^0 - \gamma_{sl}^0}{\gamma^0} \geq \frac{\pi}{2}, \tag{78}$$

with γ_{sv}^0, γ_{sl}^0, γ^0 the initial values of solid-vapor, solid-liquid, and liquid-vapor interfaces. The term «initial» means that although the adsorption process on the liquid-vapor and solid-liquid interfaces has been completed (they are fast processes), the solid-vapor interface still has its initial condition as a bare hydrophobic interface without any surfactant adsorption. At this «initial» instant of time a process of slow transfer of surfactant molecules starts from the drop onto the solid-vapor interface. Let $\Gamma_s(t)$ be the instantaneous value of surfactant adsorption onto the solid surface in front of the liquid drop on the three phase contact line, and Γ_e be the equilibrium surface density of adsorbed surfactant molecules which would eventually be reached. The driving force of the process is proportional to the difference $\Gamma_s(t)$ - Γ_e. Hence, the surfactant adsorption behaviour with time is described by

$$\frac{d\Gamma_S(t)}{dt} = \alpha\left[\Gamma_e - \Gamma_S(t)\right]$$

(79)

with the initial condition that

$$\Gamma_S(0) = 0 \text{ at } t=0$$

(80)

and $\tau_s = 1/\alpha$ is the time scale of surfactant transfer from the drop onto the solid-liquid interface at three phase contact line. Let us assume that

$$\alpha = \alpha_T \Xi \exp(\frac{-\Delta E}{kT})$$

(81)

where the prefactor α_T is determined by thermal fluctuations only; ΔE is an energy barrier for surfactant transfer from the liquid drop onto the solid-liquid interface; k, and T are Boltzman's constant and absolute temperature, respectively; Ξ is a fraction of the drop liquid-vapour interface covered with surfactant molecules. Obviously surfactant molecules position on a hydrophobic interface is "hydrophobic tails down".

We assume that transfer of surfactant molecules onto the hydrophobic solid interface takes place only from the liquid-vapour interface. It is difficult to assess the contribution of surfactant molecule transfer along the solid surface from beneath the liquid. However, our experimental data support our assumption (although they do not prove it directly). The drop surface coverage Ξ will be an increasing function of the bulk surfactant concentration inside the drop, whose maximum is reached close to the CMC. It follows from Eq. (81) that at low surfactant concentrations inside the drop τ_s should decrease with increased concentration, while above the CMC τ_s should level off and reach its lowest value. Both of these effects are observed in our experimental results (compare Fig. 8).

As the drop adopts a position according to the triangle rule, the contact angle, $\theta(t)$, is determined by the following relationship

$$\cos\theta(t) = \frac{\gamma_{sv}(t) - \gamma^0_{sl}}{\gamma^0} \tag{82}$$

where $\gamma_{sv}(t)$ is the instantaneous solid-vapour interfacial tension at the three-phase contact line. The latter dependency is determined by $\Gamma_s(t)$. According to Antonov's rule:

$$\gamma_{sv}(t) = \gamma^{\infty}_{sv}\frac{\Gamma_s(t)}{\Gamma^{\infty}} + \gamma^0_{sv}(1 - \frac{\Gamma_s(t)}{\Gamma^{\infty}}) \tag{83}$$

where γ_{sv}^{∞} is the solid-vapour interfacial tension of the surface completely covered by surfactants, and Γ^{∞} is the total number of sites available for adsorption. Hence, the final value of the contact angle can be determined from Eq. (82) as

$$\cos\theta^{\infty} = \frac{\gamma^{\infty}_{sv} - \gamma^0_{sl}}{\gamma^0} \tag{84}$$

According to Eq. (83) solid-vapour interface in front of the spreading drop changes it wettability with time: from highly hydrophobic at initial stage to partially hydrophilic at the final stage.

Using Eq. (83) in Eq. (82) yields the instantaneous contact angle

$$\cos\theta(t) = \cos\theta^0 + \lambda\frac{\Gamma_s(t)}{\Gamma^{\infty}} \tag{85}$$

where $\cos\theta^0$ is given by Eq. (78), and the positive value of λ is $\lambda = \frac{\gamma^{\infty}_{sv} - \gamma^0_{sv}}{\gamma^0}$.

Eq. (79) with initial condition (80) yields the solution

$$\Gamma_s(t) = \Gamma_e(1 - \exp(-\alpha t)) \tag{86}$$

Using (86) in Eq. (85) gives the final expression for the instantaneous contact angle

$$\cos\theta(t) = \cos\theta^0 + \lambda\frac{\Gamma_e}{\Gamma^{\infty}}(1 - \exp(-\alpha t)) \tag{87}$$

A simple geometrical consideration (Fig. 5) shows, that the radius of the wetted spot, $r(t)$, occupied by the drop can be expressed as

$$r(t) = \left(\frac{6V}{\pi}\right)^{1/3} \frac{1}{\left[\tan\frac{\theta}{2}\left(3+\tan^2\frac{\theta}{2}\right)\right]^{1/3}} \tag{88}$$

where V is the drop volume, which is supposed to remain constant and the contact angle, θ, is given by Eq. (87).

Eqs. (87), and (88) include two parameters: the dimensionless $\beta=\lambda\Gamma_e/\Gamma^\infty$ and the parameter α with dimension of inverse of time. It follows from Eq. (87) that $\beta = \cos\theta^\infty - \cos\theta^0 > 0$, where θ^∞ is the contact angle after the spreading process is completed. Both values of the contact angle, θ^0 and θ^∞, have been measured and, hence, β can be determined. Only α is used to fit the experimental data.

Let us introduce a dimensionless wetted area, $S(t)$, as

$$S(t) = \frac{r^2(t)}{\left(\frac{6V}{\pi}\right)^{2/3}} = \frac{1}{\tan^{2/3}\frac{\theta}{2}\left(3+\tan^2\frac{\theta}{2}\right)^{2/3}} = \frac{1+X}{(1-X)^{1/3}(4+2X)^{2/3}}$$

where $X=\cos\theta$, $(\cos\theta^0 \leq X \leq \cos\theta^\infty)$, which using Eq. (87) becomes $X=\cos\theta^\infty - \beta e^{-\alpha t}$.
It follows that both $dS(t)/dt$ and $dS(X)/dX$ are always positive, and the second time derivative is

$$\frac{d^2 S(t)}{dt^2} = \frac{\alpha^2}{(1-X)(2+X)}\left(\cos\theta^\infty - X\right)\frac{dS(X)}{dX}\left[-2X^2+3X\cos\theta^\infty-(2-\cos\theta^\infty)\right] \tag{89}$$

Two different situations are possible: (A) if the second derivative (89) changes sign then the spreading rate can go via a maximum/minimum value, while (B) if the second derivative (89) is always negative, the spreading rate $dS(t)/dt$ decreases with time. Case A corresponds to "high surfactant activity" $\cos\theta^\infty \geq \frac{4}{9}\left(\sqrt{10}-1\right)\approx 0.961$, while case B corresponds to "a low surfactant activity" $\cos\theta^\infty < \frac{4}{9}\left(\sqrt{10}-1\right)\approx 0.961$.

Using Eq. (84) the latter two conditions can be rewritten as $\gamma_{sv}^\infty > 0.961\gamma^0 + \gamma_{sl}^0$ at "a high surfactant activity", and $\gamma_{sv}^\infty < 0.961\gamma^0 + \gamma_{sl}^0$ at "low surfactant activity".

Under our experimental conditions we have only observed case B and, hence, we seem to have used "low surface activity" surfactants while in (Stoebe et al, 1996) "high surface activity" surfactants (or sauperspreaders) were apparently used.

3.2   Experiment
Materials.
Two types of substrate were used, a PTFE film and a polyethylene (PE) wafer. The latter substrate was prepared by crushing granules of the polyethylene composition (softening point is 100° C) between two clean glass plates under an applied pressure 1 kg/cm² at the temperature 110° C. Transparent wafers of circular section with radius 1.5 cm and thickness 0.01 cm were used.

The cleaning procedure of PTFE and PE wafers was as follows: the surfaces were rinsed with alcohol and water, then the substrates were soaked in a sulfochromic acid from 30 to 60 minutes at the temperature 50°C. The surfaces then were washed with distilled water and dried with a strong jet of nitrogen. The equilibrium macroscopic contact angles obtained were 105° and 90° for PTFE and PE substrates, respectively (for pure water droplets).

Aqueous solutions of sodium dodecyl sulfate (SDS) from Merck with weight concentration from 0.005 % up to 1 % (the CMC of the SDS is 0.2 %) were used in our spreading experiments.
Monitoring method

The time evolution of the contact line was monitored by following VCR images of drops. The images were stored using a CCD camera and a recorder at 25 frames per second. The automatic processing of images was carried out using the image-processor «Optimas». In the case of spreading over PE the initial contact angle of the drop was less than 90° and the drop was observed from above. The observed wetting area of the drop was monitored and the wetting radius was calculated. For the PTFE substrate we used a side view of the drop and, hence, the wetting radius was determined directly.

Simple mass balance estimations show that time variation of surfactant concentration inside the spreading drops can be neglected in our experiments (though it may become important in experiments of longer duration).

Water adsorption in front of the spreading drops was neglected because of hydrophobic nature of substrates used. The Peclet number in all our experiments was so small that surfactant diffusion in front of the drop was neglected.

3.3   Results and Discussion
According to our observations all the experimental drops were of spherical shape, no disturbances or instabilities were detected. Immediately after deposition the drops had a contact angle which differs slightly from the equilibrium angle of pure water on the same substrate. After a very short initial time the drops reached a position which is referred to below as "the initial" position. After that, for 1-15 seconds, depending on the SDS concentration, drops remained at the initial position. Then drops started to spread until a final value of the contact angle was reached and the spreading process was completed.

In Fig. 6 the evolution of the spreading radius of a drop over PTFE film at 0.05% SDS concentration is plotted. In Fig. 7 a similar plot is given for 0.1 % SDS concentration. In both figures, the solid lines correspond to the fitting of the experimental data by Eqs. (87) and (88), with $\tau_s = 1/\alpha$ used as a fitting parameter.

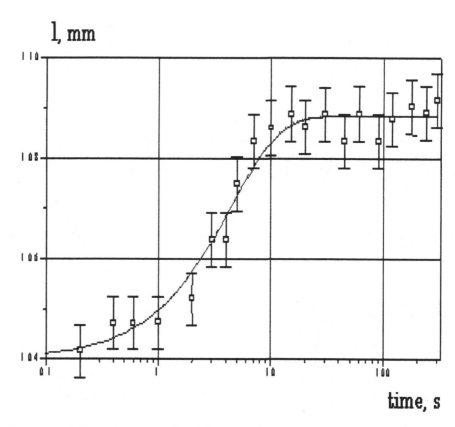

Figure 6. Time evolution of the spreading of a water drop (aqueous solution c=0.05% SDS; 2.5±0.2 µl volume) over PTFE wafer. Error bars correspond to the error limits of video evaluation of images (pixel size).

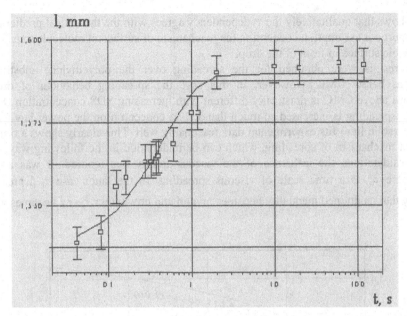

Figure 7. Time evolution of the spreading of a water drop (aqueous solution c=0.1% SDS; 2.5±0.2 µl volume) over PTFE wafer. Error bars are the same as in Fig. 6.

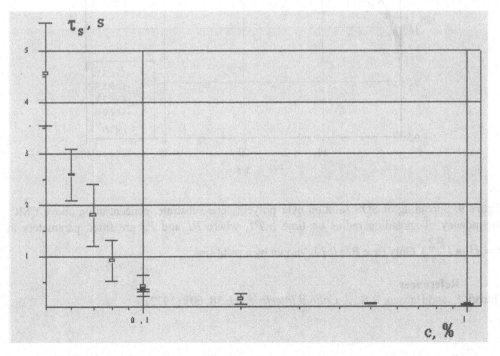

Figure 8. Fitted dependency of τ_s on surfactant concentration inside the drop (spreading over PTFE wafer). Error bars correspond to the experimental points scattering in different runs; squares are average values.

Fig. 8 shows that qualitatively the τ_s dependency agrees with the theoretical prediction and tends to support our assumption concerning the mechanism of surfactant molecule transfer onto the hydrophobic surface in front of the drop.

Similar results were obtained for the spreading over the polyethylene substrate for concentrations below CMC. However, in this case the spreading behaviour of drops at concentrations above CMC is drastically different with increasing SDS concentration (Fig. 9). The rate of a spreading is increased so much that at 1 % concentration the power law with the exponent 0.1 (solid line) fits experimental data reasonably well. This clearly shows a transition to a different mechanism of spreading, which can be understood in the following way. In our previous considerations the influence of the viscous forces was ignored. It was assumed $\tau_s >> \tau_{vis}$, where τ_{vis} is a time scale of viscous spreading. In the latter case τ_s decreases so considerably that mentioned inequality becomes invalid and now rather $\tau_s \sim \tau_{vis}$ becomes valid.

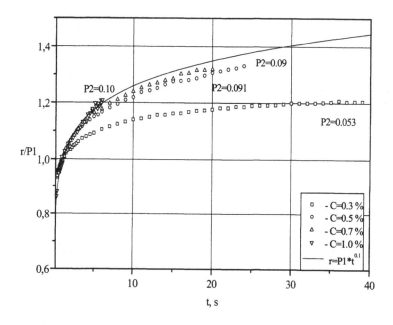

Figure 9. Spreading of SDS solution over polyethylene substrate, concentration above CMC. Dependency of spreading radius on time ($r/P1$, where $P1$ and $P2$ are fitted parameters in $r = P1 \bullet t^{P2}$). Only case $P2=0.1$ is shown by a solid line.

4   References

Ahmad, J., and Hansen, R.S., *J. Colloid Interface Sci.* **38**, 601 (1972).

Cachile, M., Cazabat, A.M., Bardon, S., Valignat, M.P., and Vandenbrouck, F. *Langmuir*, **15**, 1515 (1999).

Fraaije, J.G.E.M., and Cazabat, A.M., *J. Colloid Interface Sci.* **133**, *452* (1989).

Gaver, D.P.III, and Grotberg, J.B., *J. Fluid Mech.* **213**, 127 (1990).

Gaver, D.P.III, and Grotberg, J.B., *J. Fluid Mech.* **235**, 399 (1992).

Jensen, O.E., and Grotberg, J.B., *J. Fluid Mech.* **240**, 259 (1992).

Levich,V.G. Physicochemical Hydrodynamics. Prentice-Hall Inc., Englewood Cliffs, N.J., 1962.

Starov, V.M., Maslov, A.Yu., and Iskanderyan, G. *Colloid. J.* (English Translation), **54**, 410 (1993).

Starov, V.M., de Ryck, A., and Velarde, M.G. *J. Colloid Interface Sci.* **190**, 104 (1997).

Starov, V.M., Kosvintsev, S.R. and Velarde, M.G. *J. Colloid Interface Science*, **227**, 185 (2000).

Stoebe, T., Lin, Z., Hill, R.M., Ward, M.D., and Davis, H.T., *Langmuir*, **12**, 337 (1996); *ibid*, **13**, 7270 (1997); *ibid*, **13**, 7276 (1997).

Behaviour of Fluorochemical-Treated Fabric in Contact with Water

Ana Robert Estelrich

Pymag, S.A., Berguedà 21, Pol. Ind. Urvasa, Sta. Perpetua Mogoda, 08130 Barcelona, Spain

E-mail: arobert@pymag.com

Abstract. The introduction of fluorinated coatings in the textile industry has improved significantly the hydro- and oleophobic properties of the treated fabrics. However, these properties are notably reduced when the fabric is washed and partially recovered with heat treatment in air such as ironing. These changes of repellency are related to the modification of the chemical organisation of the fiber surface. This paper describes the chemical structure of some of the fluoropolymers used in textile and their relationship with these surface properties. A new explanation of the nature of this diminishing performance in contact with water and its recovery by heating is given.

1   Introduction

The study of the properties of solid surfaces has long involved scientific effort, both theoretical and experimental (Dhathathereyan *et al.*, 2002). Fluorinated compounds such as surface coatings have attracted especially intense research effort in recent years, and much of the motivation for such study comes from the novel physical properties of these coatings, which include high bulk modulus, reduced optical birefringence, and unusually low interfacial tension.

In the textile industry, fluorinated coatings are the most important class of water- and stain repellent finishes due to their ability to provide optimum performance in terms of both hydro- and oleophobicity. They also have additional desirable features such as fire-proofing, without impairment of the textile's permeability to air and vapour or modification of the hand of the fabric (Castelvetro *et al.*, 2001). Consequently, the importance of these coatings in clothing and as engineering textiles is growing. Introducing different monomers within the same polymer can vary and improve their performance. Today's fluorinated coating materials (usually emulsion copolymers of perfluoroalkyl-containing acrylic or methacrylic esters) are replacing the more traditional paraffin waxes and siloxane polymers (Grottenmüller, 1998; Holme, 1993).

Because fluorocarbon polymers do not adhere to most nonfluorinated surfaces, special techniques have to be adopted in order to supply sufficient adhesion between the substrate and the coating (Höpken *et al.*, 1992). The adhesion problem can be eliminated copolymerising highly fluorinated monomers with non-fluorinated ones. The polymer backbone can be tailored to provide good adhesion to the substrate, while the fluorocarbon segments are not compatible with the hydrocarbon matrix and concentrate at the coating/air-interface. The effectiveness of the surface protection depends on the coverage of the surface with fluorocarbon segments and on the degree of ordering in the surface layer.

The performance and durability of modern fabric finishes require improved laundry permanence of water and oil repellent finishes (Juhué *et al.*, 2002). However it is generally documented that water repellence decreases with extensive washing and partially recovers with ironing (Arunyadej and Mitchell, 1998). Katano *et al.* (1994) investigated the environmental effect on the surface structure in the process of generating a solid surface on the polymer films when the melted polymers are cooled either in air or in liquid (Figure 1). They found that the wettability of the films of some poly(fluoroalkyl acrylate)s and poly(fluoroalkyl methacrylate)s increases considerably when they are heated in air and then cooled in a liquid such as water. Moreover, they found that when the polymers previously treated as described above are heated again and cooled gradually in air, the surface recovers its original character.

Polymers (Katsuragawa *et al.*, 1995) containing a perfluorinated lateral chain are disordered above the melting point. When the film is cooled slowly in air, molecules are ordered so as to minimize the surface energy of the film. Therefore, for slow cooling in air, $-CF_3$ end groups of the side chains locate on the film surface. If the film is cooled rapidly in water, the side chains of molecules are arranged randomly; therefore, the presence of $-CF_3$ end groups on the surface decreases. The ordering of the formation is closely related to the surface roughness. The surface roughness of film that is rapidly cooled in water is slightly larger than that of film slowly cooled in air. The $-CF_3$ end groups are apt to be situated inside the film owing to repulsion by strongly polarized water molecules.

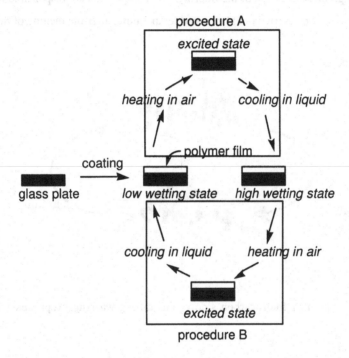

Figure 1. Schematic diagram of the procedure of heating and cooling polymer films. Reprinted with permission from Katano *et al.* Copyright 1994 American Chemical Society.

Arunyadej and Mitchell (1998) analysed the loss of surface fluorine in cotton fabrics treated with fluorocarbon by XPS data and showed that extended washing reduces the surface-fluorine concentration. Bulk-fluorine analysis of the treated fabric similarly shows a reduction in the fluorine content of 31% and 50% after one and ten washes, respectively. The effect of a hot-pressing treatment is to re-orientate the fluoropolymer as indicated by the increase in fluorine concentration. However, after 20 washes, the hot press produces a dramatic loss of surface-fluorine concentration, which is possibly due to surfactant redistribution and masking of the fluoropolymer or mechanical transfer of loosely bound fluorocarbon to the press surface.

Both states (Duschek, 2001) described are reversible and can be converted into each other (Figure 2). The high melting points of the fluorocarbon components are responsible for the fact that this conversion can only be carried out at elevated temperature, after the melting of the

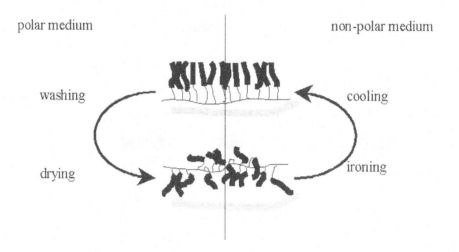

Figure 2. Re-orientation of fluorocarbon side chains during washing. Reprinted with permission from Duschek, 2001.

fluorocarbon residues and beyond the glass transition temperature of the polymer. Shorter fluorocarbon components which possess lower melting points would make the mobility at low temperature possible, however, its incompatibility with hydrocarbon residues is too low for the formation of the required boundary structure.

This means that the change in the surface characteristics of perfluoroacrylates depends on atmosphere and temperature (Hara and Katano, 2002). If a material containing perfluoroacrylates is heated and cooled in air, its surface becomes hydrophobic (Katano *et al.*, 1994). However, if it is heated and cooled in liquid, its surface becomes hydrophilic. The wetting property, particularly receding contact angles, varies in the range of 40°-110° depending on the coating composition.

Such reversible changes in repellency during washing and ironing have been explained in the literature (Sato *et al.*, 1994) by the rotation of fluoroalkyl groups into the polymer substrate to avoid direct contact with water during washing. The effect of a hot-pressing is to reorientate the fluoropolymer as indicated by the increase in fluorine concentration by XPS analyses (Juhué *et al.*, 2001). Recently (Juhué *et al.*, 2001), an alternative explanation has been proposed for limiting textile durability to laundering: this is the massive dewetting of the polymer coating which constitutes the major contribution to the diminished repellency of washed fabrics. This new concept will be extensively discussed below.

2   Surface properties

Fluorinated surfaces derive their character from the unique molecular properties associated with C-F bonding chemistry that imparts specific, unique chemistry and physics of interfaces (ACS Symposium Series, 2001). Two basic properties seem to be sought in the development of perfluorinated surfaces: low interfacial tension, and chemical resistance and durability, which means film-coating stability.

Wettability measurements can provide much useful information about the structure of polymeric surfaces (Johnson and Dettre, 1987). Similarly, polymers with well defined surfaces can give considerable insight into the interactions that control wetting.

In fact, perfluorocarbons as a class of materials exhibit the lowest possible surface tension, γ, of all organic liquids and they spread spontaneously over nearly all solid surfaces.

2.1   Physical basis for the wetting behaviour

In general, polymers have low surface energies due to the weak interaction between the different molecular groups of the polymer chain (Defay, 1966). Hence, the groups forming the polymer surface do not lack a significant interaction on their exposed surface side and the gain in surface energy from the neighbours to these groups is small (Lüning et al., 2001). In semi-fluorinated polymers, the surface energy is further lowered by reducing the polarizability of the groups that form the polymer surface. This reduces the van der Waals interaction, which generally dominates the interaction at a liquid-vapor interface.

Perfluoroalkyl substituted poly(meth)acrylates have emerged as the most widely used low surface energy polymer coatings. In these polymers, fluorocarbon chains are attached to a carboxyl group by a spacer constituted of methylene groups, resulting in a polymer structure of the form

$$\left[\begin{array}{c} R \\ | \\ C - CH_2 \\ | \\ O{=}C \\ | \\ O \\ | \\ (CH_2)_p \\ | \\ (CF_2)_q \\ | \\ F \end{array}\right]_n$$

where R is H or CH_3, p is generally 2 and q, 8.

The wetting properties of these and related polymers are believed to arise from the segregation of CF_3 groups to the surface, which have a low polarizability due to the strongly ionic character of their molecular bonds.

To explain the wetting behaviour of liquid drops on surfaces (Duschek, 2001), various scientific approaches can be used. However, the use of surface tensions or critical surface energies, which can be determined via contact angle measurement, is universally applicable and most common. High contact angles express low wetting and conversely low contact angles express good wetting (Figure 3).

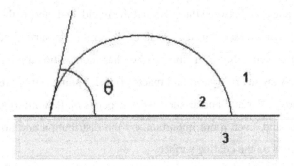

Figure 3. Liquid drop on solid substrate: 1- air, 2- liquid drop, 3- substrate

Whilst with liquids the surface tensions can be directly measured, this is not posible with solid surfaces. Therefore, a corresponding quantity for solid surfaces was derived as so-called critical surface energy γ_c, which can be extrapolated in several ways.

The connection between repellency, contact angle θ and the critical surface energy, γ_c, is described with the empirical relationship (Zisman, 1964):

$$\cos \theta = 1 + m \, (\gamma_L - \gamma_c)$$

The empirical quantity γ_c is defined as the value of γ_L (surface tension of the wetted liquid) at the intersection point of the straight line $\cos \theta$ against γ_L with the horizontal at $\cos \theta$. This extrapolation corresponds to complete surface wetting made possible by the liquid. In general, the γ_c value depends on the length, composition, branching and terminal groups of perfluoroalkyl side chains, and the crystallinity and tacticity of polymers (Pittman, 1972).

Hence, the high effectiveness of fluorinated coatings is associated with their ability to lower the critical surface tension, γ_c, of the treated surface well below that of any fluid other than a fluorocarbon (Rao and Baker, 1994). It is worth recalling that the lower the γ_c of a solid surface, the greater the contact angle, θ, between the solid and the liquid, and, hence, the lower the degree of wetting, since a liquid with surface tension γ_L can only wet a solid surface completely if $\gamma_L \leq \gamma_c$. Effectively, reducing the wettability of a fiber surface has the additional advantage of preventing liquid penetration within the micro- (yarn) and macro- (fabric) porous structure by absorption. The actual repellency to most polar and apolar substances depends on the uniformity of fiber coverage by the coating material and, even more important, by the distribution and orientation of surface-active fluorinated moieties on the coating surface.

At this point, it would be interesting to make a few statements on the wetting hysteresis phenomenon, that is the difference between advancing (θ_a) and receding (θ_r) contact angles (Game et al., 2002) (Figure 4). Static advancing and receding contact angles increase with the hydrophobicity of the substrate. Hysteresis is often present, particularly in polymers. Advancing contact angles reflect the interface between the solid "equilibrated" against the gaseous medium (often apolar air), corresponding to the interaction of the test liquid with the low energy components of the surface. Receding angles reflect the interface between the solid and the liquid corresponding to the interaction with high energy components of the surface.

The high mobility of polymers allows surface reorientation of the outermost surface moieties to minimize the free surface energy and wetting hysteresis could be ascribed to those surface reorganizations. But many other factors could contribute to hysteresis: roughness (physical heterogeneity) (Kober and Wesslen, 1994), chemical heterogeneity (Huh and Mason, 1977), diffusion or swelling of the test liquid into the material (Brandon and Marmur, 1996), or deformation of the solid surface under the liquid surface tension (Sedev et al., 1996). These different contributions may also be linked. One must therefore be vigilant when interpreting hysteresis uniquely in terms of surface reorganizations.

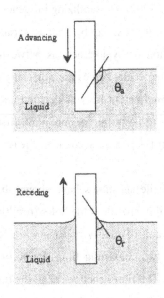

Figure 4. Determination of contact angles by the Wilhelmy plate method

With respect to the practical use of oil- and water-repellency it is pointed out that dynamic contact angles, both advancing and receding, are more important than static ones. That is, the larger the receding contact angle, θ_r, the more repellent surface we can obtain (Tetsuo, 1997). The determination of the variation of advancing and receding contact angles for surfaces of poly(perfluoroalkyl)ethyl acrylates as a function of the carbon number (n) in the side chains $C_nF_{n+2}CH_2CH_2$- shows interestingly that θ_r is very low when n is below 7. This result indicates that θ_r strongly reflects crystallinity in the fluoroalkyl side chains, since the side-chain crystallization occurs when $n \geq 7$.

2.2   Film stability

Another important matter to take into account is the film-coating stability. The evolution of thin films on solid substrates is of very great interest in many applications such as, for instance, coating or drying processes (Thiele *et al.*, 2002). Destabilizing influences leading to film rupture can arise from gradients in the surface tension caused by spatial variation of temperature or surfactant concentration fields, by evaporation or by interactions between the substrate and the film. The substrate-film interactions due to molecular forces are especially important for very thin films with a thickness smaller than one hundred nanometers. A detailed understanding of the rupture process due to molecular forces that leads to dewetting is important, first, to understand how to keep thin films stable (Kheshgi and Scriven,1991) and, second, to break thin films in a controlled manner (Mertig *et al.*, 1998).

Investigations of thin non-volatile substrates have shown that dewetting takes place in three successive phases: rupture of the film, growth of the holes (resulting in the formation of a polygonal network of straight liquid rims), and the decay of rims via a Rayleigh instability (Thiele *et al.*, 1998). It has been observed that two coexisting film rupture mechanisms occur at different film thicknesses down to below 10 nm. The film structures exhibit a distinct spatial order: holes initiated by heterogeneous nucleation at defects are randomly distributed, whereas holes resulting from spinodal dewetting, that is, spontaneous rupture under the influence of long-range molecular forces, exhibit a well developed short range order with a periodicity that corresponds to the calculated value taking polar interactions into considerations.

Hence, the stabilization of polymer thin films against dewetting is of great importance to a number of technological applications. The likelihood that films will either wet or dewet solid surfaces is related to the value of the spreading parameter S:

$$S = \gamma_B - (\gamma_L + \gamma_{LB})$$

where γ_L and γ_B are the surface tension of liquid (L) and solid substrate (B) and γ_{LB} is the LB interfacial tension (Yuang *et al.*, 1999). If S is positive, L spreads on B; if S is negative, L dewets B. Increased film stability is the result of the decrease in surface tension because the dewetting velocity is directly proportional to γ in accordance with the BMR (Brochard-Wyart/Martin/Roden) theory (Dhathathereyan *et al.*, 2002).

Polymer films can be coated onto substrates using different techniques, however, these thin films are often only metastable and will dewet when heated above their glass transition temperatures. Recently much interest has focused on the use of functional additives to control the wetting behaviour of thin polymer films. This is one of the most important techniques used nowadays to avoid dewetting in fluorocarbon films as is explained below.

3   Structure of polymers with fluorocarbone side groups

The aim of a fluorocarbon finish is to produce a surface of closely packed trifluoromethyl groups which gives the lowest possible critical surface energy (Duschek, 2001). From a chemical point of view, the generation of such structures on the fabric can be achieved by using some chemical-physical phenomena.

Perfluoroalkanes (hydrocarbon compounds in which every hydrogen atom is replaced by a fluorine atom) and alkanes (hydrocarbons) are incompatible from a certain chain length and separate spontaneously at room temperature. If a hydrocarbon residue is connected to a perfluorocarbon residue, the result is an amphiphilic molecule which is constructed from two incompatible components:

At a certain chain length (components of at least 6-8 C-atoms) (Dorset, 1990) such amphiphiles tend to self-organisation. At temperatures above 80-90°C the substances can be homogeneously melted; below 90°C a phase separation arises, i.e. the hydrocarbon residues crystallise in the proximity of hydrocarbon residues whilst the perfluoroalkyl residues only crystallise in the proximity of other perfluoroalkyl residues. During the crystallisation of such compounds in air the fluorocarbon residues arrange on the boundary surface and form a closely packed –CF_3-surface which leads to the required low critical surface energy.

Besides perfluoroalkyl monomers, acrylic acid esters with long hydrocarbon residues are commonly copolymerised in this way to realize the "principle of incompatibility" in the polymer. Further components serve to impart adhesion or cross-linking on the substrate.

The type of polymers which form these monomers are the so-called comb-shaped polymers (Volkov *et al.*, 1992). They are branched polymers that contain many side chains. For instance, each monomeric unit can contain such a chain. It is important that the length of these side groups is significantly greater than their cross-section. Only under this condition can the unique properties of comb-shaped polymers occur (for example, the autonomous behaviour of the side chains, their capacity to form layered structures, and to crystallise regardless of the main chain configuration) (Plate and Shibaev 1987).

Closely spaced side chains along the same backbone create the conditions for their interaction, which is very similar to what takes place for a group of small molecules of similar structure. On the other hand, the presence of the main chain with its inherent flexibility and intermolecular interactions leads to the dual character of the ordering and properties of the system as a whole.

The autonomous nature of the behaviour of the side chains of comb-shaped polymers makes it possible to create liquid-crystalline (LC) polymers by incorporating mesogenic groups in the side chains. Molecular models show that the general shape of the normal fluorocarbon is cylindrical. Since the van der Waals radius of fluorine is larger than that of hydrogen, the fluorine atoms attached to a planar zigzag carbon chain would be over-crowded. This slight overcrowding is relieved by rotation at each chain-bond with a slight opening of the bond angles (Bunn and Holmes, 1958) to 116°. Fluorocarbon molecules appear to be very stiff, and are unlikely to be much influenced by the weak intermolecular forces, which are similar in magnitude to those between hydrocarbon molecules. Thus, the perfluoroalkyl groups (no less than C5–C6) have an anisodiametric rod-like shape, and in this respect they are to some extent unique substances with extremely low intermolecular forces, which can form LC structures. In addition, the perfluorinated alkyl side group is a unique type of phenyl-free rigid mesogen with a strong tendency to form a smectic phase (Wang *et al.*, 1997). In comparison to LC-coil block copolymers with conventional side group mesogens, semi-fluorinated side chain block copolymers exhibit a more pronounced ability to organize in both organic solution and the solid state, as a result of their low surface energy character and strong phase separation.

3.1 Structure and phase transition of some fluorinated homo- and copolymers

Poly(alkyl methacrylate), PMAS

$$\left[\begin{array}{c} CH_3 \\ | \\ C-CH_2 \\ | \\ O=C \\ | \\ O \\ | \\ (CH_2)_m \\ | \\ H \end{array}\right]_n$$

The literature describes how homopolymers with m = 15 and m = 17 present a smectic B phase of periodicity respectively 28 and 30.4 Å (Hsieh *et al.*, 1976) with peaks at wide angles corresponding to a hexagonal lattice of parameter 4.7 Å (Shibasaki *et al.*, 1982). The pendant groups are packed perpendicular to the layers and interdigitated, which favours their crystallization.

More recently, the structure of poly(stearyl methacrylate), m = 18, has been described (Crevoisier *et al.*, 2002), with a lamellar periodicity of 29.7 Å and a crystalline Bragg peak at 4.10 Å, corresponding to a hexagonal lattice parameter of 4.7 Å (Figure 5). The size of one completely extended monomer is 23 Å, which is significantly smaller than the periodicity of 29.7 Å. The packing density is also a result of the competition between the area occupied by the pendant groups and the density of these groups along the chain: the distance between groups on the lattice (4.7 Å) is almost equal to twice their distance along the chain (2.54 Å); which means that two successive pendant groups alternate on each side of the chain and thus belong to two different layers when crystallized. DSC experiments show that the transition temperature of the smectic B (S_B) phase to an isotropic phase is 33 °C with an enthalpy of transition $\Delta H = 76$ J/g.

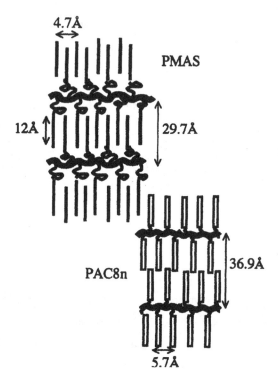

Figure 5. Schematic representation of the organization of PMAS and PAC8n. Reprinted with permission from Crevoisier *et al.* Copyright 2002 American Chemical Society.

Poly(perfluorooctylethyl acrylate), PAC8n

$$\left[\begin{array}{c} H \\ | \\ C-CH_2 \\ | \\ O=C \\ | \\ O \\ | \\ CH_2 \\ | \\ CH_2 \\ | \\ (CF_2)_8 \\ | \\ F \end{array}\right]_n$$

The organization for PAC8 is described as a smectic B with a periodicity of 33.3 Å (Mélas, 1995; Brandon and Marmur, 1996). For the same reason as in PMAS, the structure proposed is a succession of bilayers of pendant groups coming alternatively from the two different chains (Figure 5). However, contrary to PMAS, the observed periodicity is exactly twice the size of one extended monomer. This means that the groups do not interpenetrate, probably due to the large size and the rigidity of perfluorinated groups. The lattice parameter is 5.7 Å.

From DSC thermogram, optical anisotropy under polarizing optical microscope (p.o.m.) and wide-angle X-ray diffraction (w.a.x.d.) data it can be deduced that the endothermic peak at 77 °C is due to the mesomorphic-isotropic phase transition. The enthalpy of phase transition is 20.35 J/g. Hence, according to Brandon and Marmur (1996), the structure of PAC8 below 77 °C is ordered in smectic crystalline, which can be classified as smectic B (S_B) type.

Poly(perfluorooctylethyl methacrylate), PMAC8

$$
\begin{array}{c}
CH_3 \\
| \\
\left[\; C - CH_2 \; \right]_n \\
| \\
O = C \\
| \\
O \\
| \\
CH_2 \\
| \\
CH_2 \\
| \\
(CF_2)_8 \\
| \\
F
\end{array}
$$

The DSC thermogram shows three endothermic peaks (a large sharp one at 88 °C and two small ones at 43 and 110 °C), and one endothermic shoulder at ca. 68 °C. From this, from the w.a.x.d. measurements and from the optical anisotropy under p.o.m. it can be deduced that PMAC8 is in a mesomorphic state below this temperature. Consequently, the peak at 88 °C may be attributed to the mesomorphic-isotropic phase transition. The type of mesophase of PMAC8 has been classified as the smectic phase in the S_A modification. Comparison of the experimental data allow us to conclude that the side chains are arranged in a single-layer packing in the temperature range 88 – 115 °C. Furthermore, the phase transition at 88 °C can be identified as the transition of a smectic LC region with double-layer packing from the mesomorphic to an isotropic state. At temperatures below 88 °C the coexistence of single-layer and double-layer packing is presumed.

Alkyl methacrylate/perfluorooctylethyl acrylate copolymers PMAS/PAC8

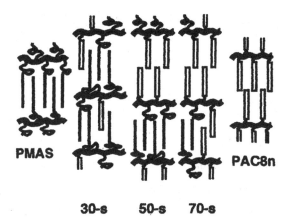

Crevoisier *et al.* (2002) have studied the copolymers bearing two types of pendant groups: one fluorinated (perfluorooctyl ethyl acrylate) group and one alkyl (equimolecular mixture of stearyl and palmityl methacrylate)

Figure 6. Schematic representation of solution copolymer organization *vs* fluorine content. The numbers (30, 50 and 70) represent the molar fractions in fluorinated pendant groups and "s" stands for statistical polymerisation. Reprinted with permission from Crevoisier *et al.* Copyright 2002 American Chemical Society.

group, varying their ratio as well as their distribution along the chain, from statistical to rather blocky. This later was achieved by changing the type of synthesis from solution to emulsion. The authors concluded that whatever the chain statistics and composition in monomers, the copolymers organize in a very structured phase, composed of a smectic whose layers are crystallized in two different lattices (Figure 6). The copolymer in solution, whose monomer distribution along the chain is statistical, leads to a smectic phase with a single periodicity roughly equal to the sum of the periodicities of the two homopolymers. On the other hand, three different periodicities appear in the blockier copolymers obtained from a synthesis in emulsion: the two homopolymer periodicities and their sum (Figure 7).

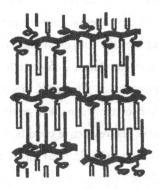

Figure 7. Schematic representation of emulsion copolymer organization. The variation of fluorine content will only modify the proportion of the two types of crystals. Reprinted with permission from Crevoisier *et al.* Copyright 2002 American Chemical Society.

4   Dewetting of polyfluoroalkyl polymers

As already mentioned in the Introduction, fluorinated polymers are used in the textile processing as hydro- and oleophobic repellents. However, it is well known that this repellency decreases significantly with washing and recovers with subsequent heat treatment. Although the reason for this change in wettability has for a long time been thought to be the transfer of fluoroalkyl groups from surface to interior of the polymer (Wakida and Li, 1993), the detailed mechanism of which having been a subject for recent studies. A dewetting of the fluorinated resin (Höpken *et al.*, 1992; Wakida and Li, 1993; Crevoisier *et al.*, 1999) has been proposed as an alternative explanation.

4.1   Dewetting mechanisms proposed

Juhué *et al.* (2001, 2002) have studied the evolution during laundering of mechanical, chemical and morphological properties using a combination of low frequency mechanical spectroscopy (LFMS), X-ray photoelectron spectroscopy (XPS), and atomic force microscopy (AFM). From the experiments of these authors two conclusions can be drawn. First, the measured stiffness of the textiles gives some direct indication of the adhesion between the fibers and the fluorinated film because a strong adhesive coating increases the stiffness of each fiber and thus the stiffness of the fabric. Second, stiffness also depends, in correlation with dewetting, on the eventual formation of some polymer bridges between cellulose fibers. Moreover, since there is no reason why textile stiffness should depend on the rotation of the fluoroalkyl groups, the mechanical spectroscopy experiments may be used to discriminate the respective contributions of the fluorocarbon rotation and the wetting/dewetting mechanism to modifications of water and oil repellency during laundering.

During washing, the evolution of a cellulose/resin/water system has to be considered, whereas during ironing it is a cellulose/resin/dry gas system. During washing, some resin dewetting from cotton fibers can be justified by thermodynamic considerations because the resin/water interface energy is high due to the hydrophobic properties of CF_2 and CF_3 entities, whereas the cellulose/water interface energy is low due to the hydrophilic character of cellulose. During ironing, there is a complete melting of the coating, which, due to the low resin viscosity at high temperature, should evolve to the thermodynamically favoured state. The resin/dry gas interface has lower energy than the cellulose/dry gas interface, which contains some polar atoms (oxygen). The more favoured state of the cellulose/resin/dry gas system should correspond to uniform spreading of the resin at the surface of the fibers.

This results lead to a new explanation for the limitation of textile durability to laundering. The authors show that the massive dewetting of the polymer coating constitutes a major contribution to the diminished repellency of washing fabrics than the fluoroalkyl groups rotation into the polymer substrate, since the rotation of the fluoroalkyl groups should not account for the observed variations of the oxygen atomic content during laundering. Obviously, pure rotations would not involve any oxygen content change.

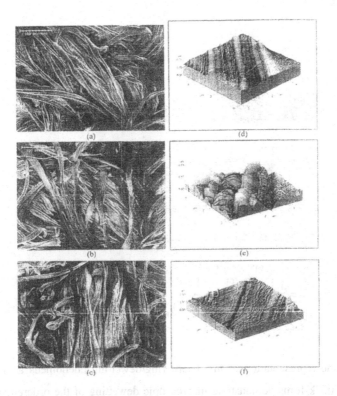

Figure 8. MEB (electron microscopy) (left) and AFM (right) investigations. Evolution of the morphology of the polymer-coated fibers during washing and ironing: (a and d) coated textile after curing, (b and e) polymer coated textile after washing, (c and f) polymer coated textile after washing and ironing. Reprinted with permission from Juhué *et al.*, 2002

Figure 8 shows the evolution in the morphology of the polymer-coated fibers during washing and ironing. After three washings at 40°C, some minor fibrillation of cotton fibers was evident from MEB (electron microscopy). Compared with a coated textile after curing (Figure 8d) there was also a strong increase in roughness (Figure 8e) observed by AFM on a smaller scale. This increase is of the order of magnitude of the polymer coating and can be attributed to a massive dewetting of the fluorinated resin during washing. According to Figure 8f, ironing allows this dewetting to be almost completely removed.

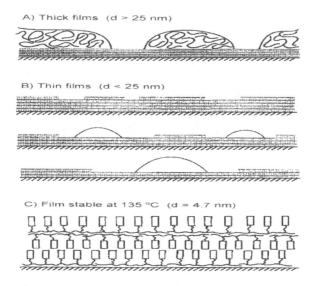

Figure 9. Schematic representation of the disordering/dewetting phenomena as a function of the film thickness: a) thick films demonstrate macroscopic dewetting of the ordered sublayers; b) the sublayers collapse via stepwise coagulation of the bilayers; c) a thin film of one and a half bilayer thickness remains stable. Reprinted with permission from Sheiko *et al.* Copyright 1996 American Chemical Society.

It has also been reported the mechanism on dewetting by increasing temperature of perfluoroalkyl methacrylate as homo- and copolymer with methyl methacrylate films. The following qualitative model has been proposed (Sheiko *et al.*, 1996). As has already been stated, perfluoroalkyl-substituted polymethacrylates have been shown to form ordered structures based on the crystallisation of the perfluoroalkyl side chains and the incompatibility of the fluorinated and the hydrocarbon segments.

The authors distinguish between three different cases for the dewetting mechanism (Figure 9). In thin films as well as in the bulk, a regular bilayer morphology is developed. Disordering into an isotropic melt occurs in two steps. A smectic mesophase is formed before the layered structure finally breaks up at elevated temperatures. This transition is characteristically effected by the interfaces in thin films on a flat substrate and, as a consequence, a peculiar self-dewetting is observed.

Thick films have been observed to become disordered above the isotropization temperature. They appear to be metastable against dewetting, so long as they do not contain large holes. However, thin films showed a peculiar stepwise dewetting where one by one bilayer collapsed and then the layers broke up irreversibly into droplets. This proceeded until a thin film with a thickness of one and a half bilayers was left. The latter remained remarkably stable even at temperatures far above the bulk disordering transitions. Interestingly, the stepwise dewetting of thin films was not observed when, instead of a perfluoroalkyl methacrylate homopolymer, a copolymer with methyl methacrylate units was used. This means that copolymers can form more stable thin films than the more regular homopolymers. This effect may correlate to the competition between the layered packing of the side chains and the coiling of the polymer backbone.

The mechanism for this peculiar layer-by-layer dewetting can be rationalized as follows. Probably the hole nucleation takes place via a "caterpillar" type relaxation of polymer backbones within one bilayer. This represents a relatively easy process as it does not require large translations of the macromolecules (Figure 10). However, hole formation should face an energetic barrier caused by the increase of the free surface area. The height of the barrier and corresponding radius of the hole depend on the hole depth and surface energies of the top layer and sublayer. Thus, only holes of a certain size will become stable and form a nucleus for further dewetting. The total dewetting process is controlled by the rate of nucleation and the growth rate of the molecules segregating into the droplets.

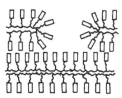

Figure 10. Hole nucleation *via* caterpillar-like retraction of the top bilayer. Reprinted with permission from Sheiko *et al.* Copyright 1996 American Chemical Society.

5 Improvement of fluorocarbon efficiency

To improve the performance and efficiency of fluorocarbons on textiles, a second product class has to be included, called boosters (Duschek, 2001) (generally urethane prepolymers on isocyanate basis), with two basic functions: 1/ chemical fixation of the fluorocarbon polymer on the textile, which leads to prolonged life of the finish and increases the efficiency, and 2/ they raise the effect level of the fluorocarbon finish by improved film formation properties and film stability. If the system of booster and fluorocarbon polymer is heated, the macroscopic thickness of the layer remains almost unchanged, as a stable network with uniform thickness is built up by the cross-linking of polymer chains. If the sample is finished without a booster, no cross-linking of the individual polymer molecules takes place. This is the reason why with fluorocarbon polymers from the basic research, the well-known phenomenon of the autophobic self-dewetting of fluorocarbon films arises (Lermann, 1997). This thinning and non-uniformity of the finish leads, then, to significantly poorer effects, even with low mechanical stress to the textile. In addition, such finishes clearly show a lower wash permanence.

6 Conclusions

It has been shown that fluorinated polymers present a complex structure with a transition between smectic lamellar to isotropic phase when the temperature is increased (Bunn and Holmes, 1958). Through the side-chain ordering, the smectic structure brings hardness and non-wettability. In contrast, in the isotropic phase, the presence of the backbone, which connects the side chains together, allows for a strong dissipation that leads to an ability to slow down the dynamics of wetting. On each side of the phase transition, a different aspect of the hybrid macromolecule becomes predominant and imprints its behaviour onto the system. This effect can be used to design versatile materials with highly flexible properties that vary with temperature. Moreover, the transition temperature can be tuned by changing the copolymer composition, which may be useful for different types of applications. For instance, the addition of specific co-monomers, which increase the polymer melting temperature above the washing temperature, improves the behaviour of fabrics during laundering.

In conclusion, it seems that improvement of fluorocarbon performance may be possible by verifying which monomers could be useful and effective for the control or, indeed, the elimination of dewetting for thin films of FC polymers.

References

ACS Symposium Series 2001, 787 (Fluorinated Surfaces, Coatings and Films), 1

Arunyadej S.,and Mitchell R., *J. Text. Inst.* 1998, **89 Part 1**, 696

Brandon S., and Marmur A., *J. Colloid Interface Sci.* 1996, **183**, 351

Bunn C. W., and Holmes D. R., *Disc. Faraday Soc.* 1958, **25**, 95

Castelvetro V., Francini G., Ciardelli G., and Ceccato M., *Textile Research Journal* 2001, **71/5**, 399

Crevoisier G., Fabre P., Corpart J. M., and Leibler L., *Science* 1999, **285**, 1246

Crevoisier G., Fabre P., Leibler L., Tencé-Girault S., and Corpart J. M., *Macromolecules* 2002, **35(10)**, 3880

Defay R., Surface Tension and Adsorption, Wiley, New York, 1966

Dhathathereyan A., Baskar G., Ramasami T., and Juhué D., *Langmuir* 2002, **18(12)**, 4704

Dorset D.L., *Macromolecules* 1990, **23**, 894

Duschek G., *Melliand Textilberichte* 2001, **7-8**, 604

Game Ph., Sage D., and Chapel J. P., *Macromolecules* 2002, **35**, 917

Grottenmüller R., Fluorocarbons-An Innovative Aid to the Finishing of Textiles, *Melliand Int.* 1998, **4**, 278

Hara K., and Katano Y., *Japanese J. App. Phys., part 2: Letters* 2002, **41(4B)**, L490

Holme, I., New Developments in the Chemical Finishing of Textiles, *J. Textile Inst.* 1993, **84**, 520

Höpken J., Sheiko S., Czech J., and Möller M., Polymer Preprints (American Chemical Society, Division of Polymer Chemistry) 1992, 33(1), 937

Hsieh H. W. S., Post, B., and Morawetz, H., *J. Polym. Sci., Polym. Phys. Ed.* 1976, **14**, 1241

Huh C., and Mason S. G., *J. Colloid Interface Sci.* 1977, **60**, 11

Johnson R. E. Jr., and Dettre R. H., Polymer Preprints (American Chemical Society, Division of Polymer Chemistry) 1987, 28(2), 48

Juhué D., Gayon A. C., Corpart J. M., Charret N., Cavaillé J. Y., and Perriat P., Fluorine in Coatings IV, paper 22, 2001

Juhué D., Gayon A. C., Corpart J. M., Quet C., Delichère P., Charret N., David L., Cavaillé J. Y., and Perriat P., *Textile Research Journal* 2002, **72(9)**, 832

Katano Y., Tomono H., and Nakajima T., *Macromolecules* 1994, 27, 2342

Katsuragawa T., Chiba E., Okada K., Tani K., and Tomono H., *Jpn. J. Appl. Phys.* 1995, **34**, 649

Kheshgi H. S., and Scriven L. E., *Chem. Eng. Sci.* 1991, **46**, 519

Kober M., and Wesslen B., *J. App. Polym. Sci.* 1994, **54**, 793

Lermann E., Dissertation Kapitel 6, Univerität Ulm 1997

Lüning J., Yoon D.Y., and Stöhr J., *J. Electron Spectroscopy and Related Phenomena* 2001, **121**, 265

Mélas M., Thèse de Doctorat, Université de Montpellier II, 1995

Mertig M., Thiele U., Bradt J., Klemm D., and Pompe W., *Appl. Phys.* 1998, **A66**, S565

Pittman A.G., "Fluoropolymers", Wall, L.A. ed., Wiley-Interscience, New York, 1972, pp. 419-49

Plate N. A., and Shibaev V. P., 'Comb-Shaped Polymers and Liquid Crystals', Plenum Press, New York, 1987

Rao N. S., and Baker B.E., Textile Finishing and Fluorosurfactants in Organofluorine Chemistry: Principles and Commercial Applications, R.E. Banks et al., Eds. Plenum press, NY, 1994, pp. 321-338

Sato Y., Wakida T., Tokino S., Niu S., Ueda M., Mizushima H., and Takekoshi S., *Textile Research Journal* 1994, **64(6)**, 316

Sedev R. V., Petrov J. G., and Neumann A. W., *J. Colloid Interface Sci.* 1996, **180**, 36

Sheiko S., Lermann E., and Möller M., *Langmuir* 1996, **12**, 4015

Shibasaki Y., and Fukuda K., *Therm. Anal., Proc. Int. Conf. 7th* 1982, **2**, 1517

Tetsuo S., Modern Fluoropolymers, Chapter 6 (1997 John Wiley & Sons Ltd.)

Thiele U., Mertig M., and Pompe W., *Phy. Rev. Lett.* 1998, **80(13)**, 2869

Thiele U., Neuffer K., Pomeau Y., and Velarde M. G., *Colloids and Surfaces* A 2002, **206**, 135

Volkov V., Platé N., Takahara A., Kajiyama T., Amaya N., and Murata Y., *Polymer* 1992, **33**, 1316

Wakida T., and Li H., *J. Soc. Dyers Color.* 1993, **109**, 292

Wang J., Mao G., Ober C. K., and Kramer E. J., *Macromolecules* 1997, *30*, 1906

Yuang C., Ouyang M., and Koberstein J. T., *Macromolecules* 1999, **32**, 2329

Zisman W.A., "Contact Angle, Wettability, and Adhesion", Good, R F. (ed.), Adv. Chem. Ser. 43, 1964, 1, Am.Chem. Soc., Washington DC

Printed in the United States
By Bookmasters